Genes, Peoples, and Languages

Genes, Peoples, and Languages

Luigi Luca Cavalli-Sforza

Translated by Mark Seielstad

ALLEN LANE
THE PENGUIN PRESS

ALLEN LANE
THE PENGUIN PRESS

Published by the Penguin Group
Penguin Books Ltd, 27 Wrights Lane, London W8 5TZ, England
Penguin Putnam Inc., 375 Hudson Street, New York, New York 10014, USA
Penguin Books Australia Ltd, Ringwood, Victoria, Australia
Penguin Books Canada Ltd, 10 Alcorn Avenue, Toronto, Ontario, Canada M4V 3B2
Penguin Books India (P) Ltd, 11, Community Centre, Panchsheel Park, New Delhi – 110 017, India
Penguin Books (NZ) Ltd, Private Bag 102902, NSMC, Auckland, New Zealand
Penguin Books (South Africa) (Pty) Ltd, 5 Watkins Street, Denver Ext 4, Johannesburg 2094, South Africa

Penguin Books Ltd, Registered Offices: Harmondsworth, Middlesex, England

Originally published under the title *Gènes, peuples et langues*
First published in English under the title *Genes, Peoples and Languages*
by North Point Press, a division of Farrar, Straus & Giroux 2000
First published in Great Britain by Allen Lane The Penguin Press 2000
3

Printed and bound in Great Britain by The Bath Press, Bath
Cover repro and printing by Concise Cover Printers

A CIP catalogue record for this book is available from the British Library

ISBN 0–713–99486–X

CONTENTS

Preface

This book surveys the research on human evolution from the many different fields of study that contribute to our knowledge. It is a history of the last hundred thousand years, relying on archeology, genetics, and linguistics. Happily, these three disciplines are now generating many new data and insights. All of them can be expected to converge toward a common story, and behind them must lie a single history. Singly, each approach has many lacunae, but hopefully their synthesis can help to fill the gaps. Other sciences—cultural anthropology, demography, economy, ecology, sociology—are joining in the effort, and are justly becoming pillars of interpretation.

It would be impossible to communicate the conclusions about human history and the causes of human evolution if one had to rely on the jargons of such diverse disciplines. Scientific terminology insures precision and increases the speed of communication among specialists, but it creates a barrier between them and the general public. I have tried to restrict my use of jargon to a minimum, and I have also attempted to explain terms and methods unknown to the general public. The response to foreign editions of this book (in French, Italian, Spanish, and German) indicates that most readers have no difficulty in following the science presented here, and can appreciate multidisciplinarity.

To some, history (including evolution) is not a science, because its results cannot be replicated and thus cannot be tested by the experimental method. But studying the same phenomenon from many different angles, from many disciplines, each of which supplies independent facts, has the value of largely independent repetition. This makes the multidisciplinary approach indispensable.

An important conclusion that emerges from this work is that human genetic evolution has been heavily affected by technological innovations and by cultural change, in general. Culture, meaning the accumulation of knowledge over generations, is the main difference between humans and other animals (the difference is one of degree, because animals, too, learn during their lives and transmit knowledge to future generations). Cultural transmission is thus an important object of study, one that has been dramatically neglected. Chapter 6 is devoted to it.

The subject of this book has significant implications for important social problems. It explains, among other things, why racism is fallacious. Genetics is instrumental in shaping us, but so, too, are the cultural, social, and physical environments in which we live. The main genetic differences are between individuals and not between populations, or so-called "races." Differences of genetic origin among the latter are not only small (rapidly becoming even smaller with the recent acceleration of transportation, and both migratory and cultural exchange) but also superficial, attributable mostly to responses to the different climates in which we live. Moreover, there are serious difficulties in distinguishing between genetic and cultural differences, between nature and nurture.

My greatest hope is that the reader experiences the same intellectual pleasure I have with each expected and unexpected finding, uncovering so many points of agreement among disciplines that have been kept carefully separate for so long.

Acknowledgments

This book owes much to many people. It was conceived when I was invited in 1981 and 1989 to give a series of lectures at the Collège de France. The "Collège" is a wonderful institution created by Francis I to counteract the arrogance and backwardness of the Sorbonne, and to exemplify a center of excellence. Jacques Ruffié's arrangements motivated me to write my lectures, and gave me the wonderful chance to twice spend a month in Paris in the spring, delivering them. In 1994, Odile Jacob expressed interest in publishing the lectures in a new series of books based on courses at the Collège, and so I rewrote my lectures from scratch for the third time. The Italian edition was a fourth opportunity. My former student Mark Seielstad was not deterred by significant differences between the French and the Italian editions, but used them both in order to improve the English translation, which he completed while in the throes of preparing his Ph.D. dissertation at Harvard. The need to revise the English edition was also a new and irresistible temptation for me to update and modify the book again. This fifth version has been edited with great skill and attention to clarity, rigor, and accuracy by Ethan Nosowsky. I am also grateful to Phyllis Mayberg for her help in preparing the manuscript and to Brian Blanchfield for ushering it to press.

Collaboration with experts from other fields is essential when taking a multidisciplinary approach. I am indebted to many friends and colleagues who, over the last fifty years, have helped lay the foundation for the scientific work described in this book. As an expression of gratitude, I'd like to summarize the major collaborations of these many decades.

I began as a geneticist researching bacteria in the forties. In the fifties, when I was teaching part-time in the University of Parma, Italy, I gradually shifted focus to human population genetics. My main research at Parma was the study of the role of chance in evolution, a subject that was neglected at the time. The opportunity to give a clear, quantitative answer to the problem was offered by the demographic records of the last three centuries or more in a region where the population density varied enormously, very high in the fertile plains, very low in the mountains. Sizes of villages and migration among them could be estimated from the available parish books. When there are few parents to pass genes from one generation to the next, chance will cause important fluctuations in the frequencies of types of genes found in different villages. The effect of chance in evolution is called "drift," a somewhat misleading name, because the word practically has the opposite meaning in other sciences. This research made it possible to predict the variation between villages due to drift on the basis of their demography, and compare it with local genetic variation. This work and a parallel study of consanguinity data from bishopric archives would not have been possible without the advice, information, and help of Antonio Moroni, a Catholic priest, then my student and now professor of ecology at Parma, and Franco Conterio, then postdoctoral fellow and now professor of Anthropology at Parma.

In the sixties I moved to the University of Pavia, Italy, and started generating methods for reconstructing evolutionary trees of human genetic data, in collaboration with Anthony Edwards, now at Gonville and Caius College of Cambridge. Afterward I researched African Pygmies in many expeditions between 1976 and 1985. This work profited greatly from my collaboration with Marcello Siniscalco, then professor at Leyden, as well as with anthropologists Colin Turnbull, now, unfortunately, deceased, and Barry Hewlett, now a professor in Vancouver, Washington. Our work was the subject of a book I edited, published in 1986, called *African Pygmies.*

It soon became clear that human population genetics research, touching as it does on many disciplines, could prosper only with the help of scientists from other disciplines. I moved to Stanford in 1971 and collaborated with archeologist Albert Ammerman, now at Colgate University, on the problem of whether, in the expansion of Neolithic agriculture from the Middle East to Europe, the farming technique or the farmers themselves diffused northwest from the place of origin. The study of genetic geography was begun in 1977 with the collaboration of Paolo Menozzi and Alberto Piazza, now professors of ecology at Parma and of human genetics at Turin respectively, with the purpose of offering a solution to that problem, and we provided a key to the answer. Eventually this approach was extended to the rest of the world and led to the publication of our *History and Geography of Human Genes* in 1994 with Princeton University Press. It is the source for the majority of claims made in the first five chapters of this book, and is referred to by the acronym HGHG.

In the seventies and early eighties I dedicated much time to the study of cultural evolution, mostly as a response to the personal and profound interest I developed after observing African Pygmies. Researching cultural transmission and evolution, I gainfully collaborated with Marcus Feldman, professor of biology at Stanford. Applications to linguistic evolution were made possible thanks to exchanges with linguists from the Bay Area: Bill Wang, from Berkeley, Joseph Greenberg and Merritt Ruhlen at Stanford.

At the end of the seventies and beginning of the eighties, with the seminal work of Y. W. Kan, David Botstein, Ronald Davis, Mark Skolnick, and Ray White, the promise of chemical analysis of DNA—the material of heredity—started to become reality. Until that time, genes—the units of inheritance—were interpreted only through their products, mostly proteins. From then on it became possible, and eventually much easier, to study variation on DNA directly.

Mitochondrial DNA, a small organ that is present in every cell and transmitted to the progeny by the mother, was one of the first objects of study. We began this work with Doug Wallace and his students. In the hands of the late Allan Wilson of Berkeley, it gave the first important evidence that modern humans appeared in Africa and from there spread to the rest of the world. Now the purposes of our research are also served by studying Y chromosomes, found only in males and transmitted from father to son. I was blessed by a stroke of good luck when Peter Underhill of my laboratory and Peter Oefner of Ron Davis's laboratory together developed a superior technique of detecting DNA variation. The new genealogy of Y chromosome variants they developed will enormously aid our understanding of the history of the evolution of modern humans. This research is now progressing rapidly.

The results already at hand promise to generate a clear picture of the expansions and migrations out of Africa that made modern humans look the way they do today. It would seem that these events were much more recent than has been thought. It is impossible to generate much diversity in such a short period of time, which convinces us once and for all that the superficial racial differences we perceive between people from different continents are just that.

Genes, Peoples, and Languages

CHAPTER 1

Genes and History

The Pride of an Emperor

Dante Alighieri's reputation as the grand master of Italian literature has eclipsed all the Italian poets and writers who followed him. Nevertheless, Dante was not the only great Italian poet. There were others, such as Petrarch, Ariosto, and Leopardi. The latter is perhaps the least well-known outside Italy, although he was not only a talented poet but also a remarkable philosopher.

I recently reread his play *Copernicus,* which I still find relevant and insightful. The characters include the Sun, the First and Last Hours of the Day, and Copernicus. In the opening scene, the Sun confides to the First Hour that he is tired of revolving around the Earth each day, and demands that the Earth shoulder some of the burden. The First Hour, alarmed by this prospect, points out that the Sun's retirement would create havoc. But the Sun is adamant and insists on informing Earth's philosophers of the impending change since he believes they can convince humans of anything—good or bad. By the second scene, the Sun has delivered on his threat. Copernicus, surprised by the Sun's failure to rise, sets about investigating the cause. His search quickly ends when he and the Last Hour are summoned to hear the Sun's proposal: the Earth must renounce her position at the center of the Universe and instead revolve around the Sun. Copernicus notes that even philosophers would have difficulty convincing the Earth of that. Moreover, the Earth and her inhabitants have grown accustomed to their position at the center of the Universe and have developed the "pride of an emperor." A change of such magnitude would have not only physical but also social and philosophical consequences. The most basic assumptions about human life would be overturned. But the Sun is insistent that life will go on, that all the barons, dukes, and emperors will continue to believe in their importance, and that their power won't be weakened in the least. Copernicus offers further objections: a galactic revolution could begin—the other planets may assert that they want the same rights to centrality as the Earth had. Even the stars would protest. In the end, the Sun might lose all importance and be forced to find another orbit. But the Sun desires only rest and counters Copernicus's final fear—that he will be burned as a heretic—by telling him he can avoid such a fate by dedicating his book to the Pope.

In writing about Copernicus, Leopardi had the benefit of living several centuries after him. He knew what had happened to Copernicus, Giordano Bruno, and Galileo. But we do not have Leopardi's advantage when considering the scientific issues of our day. Any current theories may be modified or even destroyed at any moment. In fact, science progresses because every hypothesis can be confirmed or rejected by others. The great number of conditionals we use in our scientific prose underscore this truth. While correcting the translation of one of my books, I was terrified to see that all my conditionals had been changed to indicatives—my safeguards had been eliminated. When we write papers for scientific journals, we know that many statements cannot be supported in their entirety. This seems strange to the public: isn't science infallible? In the end, only religion claims to deliver certainty. In other words, faith alone is immune from doubt, although few believers seem troubled by the fact that each religion offers different answers. Mathematics may be the only exception in the sciences that leaves no room for skepticism. But, if mathematical results are exact as no empirical law could ever be, philosophers have discovered they are not absolutely novel—instead, they are tautological.

Copernicus also reminded me of our attitudes about race and racism. Each population believes that it is the best in the world. With few exceptions, people love the microcosm into which they are born and don't want to leave it. For Whites, the greatest civilization is European; the best race is White (French in France and English in England). But what do the Chinese think? And the Japanese? Wouldn't most of today's recent immigrants return to their country if they could find a decent way of life there?

It is also true, as Leopardi observed, that the more things change, the more they stay the same. Noble or economically powerful families come and go—there is an increasingly rapid turnover of power—but power structures change very little. The Roman Empire lasted longer than many others in Europe, but it spanned only five centuries. It was similar in size to the Inca Empire, which lasted a little more than a century. Before the Roman Empire, several maritime powers—the Greeks, Phoenicians, and Carthaginians—colonized the Mediterranean coast. At the same time, the European interior saw Celtic princes establish control over most of Europe. During the second half of the first millennium B.C., the Celtic and maritime fiefdoms were each united by commercial, linguistic, and cultural ties, but were politically fragmented.

Ultimately, they would all fall to the Romans. The Romans built the first politically united culture in Europe, but it eventually fell to "barbarian" invaders from the East. The barbarians flourished, and only the eastern part of the Roman Empire—the Byzantine Empire—was to survive into the Middle Ages. In the west, Charlemagne founded the Holy Roman Empire in A.D. 800, the culmination of Frankish political development. France, Germany, and parts of Italy and Spain were briefly reunited. After A.D. 1000, Frankish power passed to Germany and, in part, to the Pope, although the

Papacy and the Empire were often in conflict. The Holy Roman Empire ceased to have any political importance by the fourteenth century, although Austrian emperors continued to take the title of Holy Roman Emperor until 1806. Several European states were formed or consolidated between 1000 and 1500. Although wars among them were frequent, none was able to conquer much of Europe before Napoleon. With the development of seaworthy ships, the armies and navies of Europeans attempted to extend their hegemony to the rest of the world, competing for national riches on other continents. The Portuguese, Spanish, English, Dutch, French, and Russians established overseas empires which would endure into the twentieth century, but in all of European history, not a single empire has lasted for more than five centuries. Napoleon rapidly conquered continental Europe, but his rule lasted for fewer than ten years.

The Chinese Empire began in the third century B.C. and endured many vicissitudes under myriad dynasties, none of which lasted for more than four centuries. After several difficult periods, China fell to the Mongols in the thirteenth century. One hundred years later, the Ming restored Chinese dominance for three centuries. Then another foreign dynasty, the Qing, ruled for several centuries into the twentieth. The same pattern is found on every continent or subcontinent.

National pride is always more fervent in successful times. When a people feels strong, it is easier to say, "We are the best." However, power can have rather unusual origins. The wise decisions and shrewd political acts of a few leaders or small groups often produce enduring states. Even cruel regimes can sometimes succeed in introducing prosperous periods. The rise to political power frequently requires violence, which is not always physical. Favorable external circumstances can also help maintain stability, if only temporarily. Politicians who wield their power responsibly are difficult to replace with equally capable successors. During happy and prosperous years, people can convince themselves that their success is due to their excellent qualities, the intrinsic characteristics of their "race" that make them great. The illusion of immortality ignores all the lessons of history. The self-critic is rare and tends to be absent or has no listeners when things are going well.

Perhaps Claude Lévi-Strauss most succinctly defined racism as the belief that one race (usually, though not always, one's own) is biologically superior—that superior genes, chromosomes, DNA put it at an advantage over all others. This is America's situation now. It is no coincidence that you must first dial the number one when calling the United States from abroad.

At any particular moment, a single people may be dominant despite the many countries that have been before, or will be soon. Of course, it is not necessary to *be* superior to be convinced that one is. Even a limited success can demonstrate power to others. Many believe such dominance is determined by biology.

Other Sources of Racism

Almost any society can find a good reason to consider itself predominant, at least in a particular activity. A simple claim to competence in any sphere—be it painting, football, chess, or cooking—is often sufficient to imbue a people with exaggerated importance.

One's daily routine, which is subject to both individual and cultural influences, is filled with superficial comparison of one's own habits with foreign, often significantly different, habits. Even if we do not know the sources of these differences, the simple fact that they exist can be enough to inspire fear or hatred. Human nature does not welcome change, even when we're dissatisfied with things as they are. Perhaps this devotion to habit and fear of melioration encourage a conservatism that could lead to racism.

There are unquestionable differences among peoples and nations. Language, skin color, tastes (especially in food), and greeting all differ among cultures and lead us to believe that others are essentially not like us. We typically conclude that our ways are the best, and too bad for the others. To the Greeks, all those who did not speak Greek were barbarians. Of course, when a person is unsatisfied with life in his home country and migrates, he might more easily tolerate uncertainties and strange living conditions in another region or continent. He might even accept the necessity of learning new things. But in general, he prefers the cocoon in which he was born, terrified of discarding what is familiar.

Many other factors nourish racist sentiments. One of the most important is the desire to project one's unhappiness onto another. Everyone knows that self-alienation in modern society is often a very serious cause of irritation and angst. These feelings can arise from the fear of unemployment, being forced to perform inhumane work, the reality and experience of poverty and injustice, and the feeling of powerlessness which often results from the jealous observation that vast wealth is possible only for the very few. Everyone, even those who feel victimized by their superiors, can assume authority over those lower on the social ladder. The poor can always find somebody poorer.

Because of all these factors, racism is widespread. It is less apparent during times of peace and civil order. But hostilities about mass immigration from poor countries exacerbate it.

Is There a Scientific Basis for Racism?

Racism should be condemned because its effects are pernicious. It is criticized by virtually every modern religion and ethical system. However, can we exclude the possibility that a superior race exists, or that socially important, inherited differences between the races can be found? There are certain obvious differences between human groups for traits that depend to some extent on genes: skin color, eye shape, hair type, facial form, and body shape. Will these and other traits provide a scientific justification for racism? Do other differences exist that might?

We must first define the nature of the variation to be studied. Doing so helps us to understand what we mean by race, to decide which groups we should examine and what racial differences may tell us.

Biological and Cultural Variation

We must note that most people do not distinguish between biological and cultural heredity. It is often difficult to recognize which is which. Sometimes the cause of racial difference is biological (in which case we call it genetic, meaning that it comes with your DNA); sometimes it is behavioral, learned from someone else (these are cultural causes); and sometimes both factors are involved. Genetically determined traits are very stable over time, unlike socially determined or learned behavior, which can change very rapidly. As I said above, there are clear biological differences between populations in the visual characteristics that we use to classify the races. If these genetic differences were found to be genuinely important and could support the sense of superiority that one people can have over another, then racism is justified—at least formally. I find this genetic or biological definition of racism more satisfactory than others. Some would extend the domain of racist judgments to include any difference between groups, even the most superficial cultural characteristics. The only advantage of this broader definition is that it sidesteps the difficulty of determining whether certain traits have a genetic component or not. But it does not seem appropriate to speak of racism when one person resents another's loud voice, noisy eating habits, taste in dress, or difficulties with correct pronunciation. This type of intolerance, which is rather common in certain countries or social classes, seems much easier to correct and control through education than is true racism.

Visible and Hidden Variation

The racial differences that impressed our ancestors and that continue to bother many people today include skin color, eye shape, hair type, body and facial form—in short, the traits that often allow us to determine a person's origin in a single glance. Ignoring admixture, it is fairly easy to recognize a European, an African, and an

Asian, to mention those standard types with which we are most familiar. Many of these characteristics—almost homogeneous on a particular continent—give us the impression that "pure" races exist, and that the differences between them are pronounced. These traits are at least partly genetically determined. Skin color and body size are less subject to genetic influence since they are also affected by exposure to the sun and diet, but there is always a hereditary component that can be quite important.

These characteristics influence us a lot, because we recognize them easily. What causes them? It is almost certain they evolved in the most recent period of human evolution, when "modern" humans, or early humans practically undistinguishable from ourselves, first appeared in Africa, grew in numbers, and began to expand to the other continents. Evidence and details will be discussed later. What interests us here is that this diaspora of Africans to the rest of the world exposed them to a great variety of environments: from hot and humid or hot and dry environments (to which they were already accustomed) to temperate and cold ones, including the coldest ones of the world, as in Siberia. We can go through some of the steps that this entailed.

1. Exposure to a new environment inevitably causes an adaptation to it. In the 50,000–100,000 years since the African diaspora, there has been an opportunity for substantial adaptation, both cultural and biological. We can see traces of the latter in skin color and in size and shape of the nose, eyes, head, and body. One can say that each ethnic group has been genetically engineered under the influence of the environments where it settled. Black skin color protects those who live near the equator from burning under the sun's ultraviolet radiation, which can also lead to deadly skin cancers. The dairy-poor diet of European farmers, based almost entirely on cereals that lack ready-made vitamin D, might have left them vulnerable to rickets (our milk still has to be enriched with this vitamin). But they were able to survive at the higher latitudes to which they migrated from the Middle East because the essential vitamin can be produced, with the aid of sunlight, from precursor molecules found in cereals. For this Europeans have developed the whiteness of their

skin, which the sun's ultraviolet radiation can penetrate to transform these precursors into vitamin D. It is not without reason that Europeans have, on average, whiter skin the further north they are born.

The size and shape of the body are adapted to temperature and humidity. In hot and humid climates, like tropical forests, it is advantageous to be short since there is greater surface area for the evaporation of sweat compared to the body's volume. A smaller body also uses less energy and produces less heat. Frizzy hair allows sweat to remain on the scalp longer and results in greater cooling. With these adaptations, the risk of overheating in tropical climates is diminished. Populations living in tropical forests generally are short, Pygmies being the extreme example. The face and body of the Mongols, on the other hand, result from adaptations to the bitter cold of Siberia. The body, and particularly the head, tends to be round, increasing body volume. The evaporative surface area of the skin is thus reduced relative to body volume, and less heat is lost. The nose is small and less likely to freeze, and the nostrils are narrow, warming the air before it reaches the lungs. Eyes are protected from the cold Siberian air by fatty folds of skin. These eyes are often considered beautiful, and Charles Darwin wondered if racial differences might not result from the particular tastes of individuals. He called the idea that mates were chosen for their attractive quality "sexual selection." It is very likely that some characteristics undergo sexual selection—eye color and shape, for example. But the shape of Asian eyes is not appreciated only in Asia. If it is admired elsewhere, why is it not found in other parts of the world? Of course it is also characteristic of the Bushmen of southern Africa, and other Africans have slanted eyes. It probably diffused by sexual selection from northeastern Asia to Southeast Asia, where it is not at all cold. It is also possible that the trait may have originated more than once in the course of human evolution. If it first appears that climatic factors were most important in the creation of racial differences, we should not neglect sexual selection as a possible side explanation. Unfortunately, the genetic bases for these adaptations are not known; each of these traits is very complex. Considerable local variation in tastes further complicates the matter.

2. There is little climatic variation in the area where a particular population lives, but there are significant variations between the climates of the Earth. Therefore, adaptive reactions to climate must generate groups that are genetically homogeneous in an area that is climatically homogeneous, and groups that are very different in areas with different climates.

We could ask if sufficient time has passed since the settling of the continents to produce these biological adaptations. The selection intensity has been very strong, so the answer is probably yes. We could note in this regard that the Ashkenazi Jews who have lived in central and eastern Europe for at least 2,000 years have much lighter skin than the Sephardi Jews who have lived on the Mediterranean perimeter for at least the same length of time. This could be an example of natural selection, but it might also result from genetic exchange with neighboring populations. Some available genetic information favors the second interpretation, but better genetic data are desirable before we can exclude the influence of natural selection.

3. Adaptations to climate primarily affect surface characteristics. The interface between the interior and exterior plays the biggest part in the exchange of heat from the interior to the exterior and vice versa. A simple metaphor can help explain this statement: if you want to decrease the cost of heating your house in the winter, or cooling it in the summer, you must increase the house's insulation so that the thermal flow between the inside and outside is minimal. Thus, body surface has been largely modified to adapt different people to different environments.

4. We can see only the body's surface, as affected by climate, which distinguishes one relatively homogeneous population from another. We are therefore misled into thinking that races are "pure" (meaning homogeneous) and very different, one from the other. It is difficult to find another reason to explain the enthusiasm of nineteenth-century philosophers and political scientists like Gobineau and his followers for maintaining "racial purity." These men were convinced that the success of whites was due to their racial supremacy. Because only visible traits could be studied then, it was

not absurd to imagine that pure races existed. But today we know that they do not, and that they are practically impossible to create. To achieve even partial "purity" (that is, a genetic homogeneity that is never achieved spontaneously in populations of higher animals) would require at least twenty generations of "inbreeding" (e.g., by brother-sister or parent-children matings repeated many times). Such inbreeding would have severe consequences for the health and fertility of the children, and we can be sure that such an extreme inbreeding process has never been attempted in our history, with a few minor and partial exceptions.

In more recent times, the careful genetic study of hidden variation, unrelated to climate, has confirmed that homogeneous races do not exist. It is not only true that racial purity does not exist in nature: it is entirely unachievable, and would not be desirable. It is true, however, that "cloning," which is now a reality in animals not very remote from us, can generate "pure" races. Identical twins are examples of living human clones. But creating human races artificially by cloning would have potentially very dangerous consequences, both biologically and socially.

We shall also see that the variation between races, defined by their continent of origin or other criteria, is statistically small despite the characteristics that influence our perception that races are different and pure. That perception is truly superficial—being limited to the body surface, which is determined by climate. Most likely only a small bunch of genes are responsible, and little significance is attached to them, especially since we are progressively developing a totally artificial climate.

Hidden Variation: Genetic Polymorphisms

The ABO blood group was the first example of an invisible and completely hereditary trait. Discovered at the beginning of the century, it has been the subject of numerous studies, because the matching of blood types is essential for successful blood transfusions. There are three major forms of the gene (also called "alleles"): A, B, and O, and they are strictly hereditary. An individual can have one of four possible blood types: O, A, B, and AB.

Although it is not truly essential for the understanding of what follows, it is difficult to resist the opportunity of mentioning at this point a basic rule of inheritance: each of us receives one allele from each parent—one from the father and one from the mother. Therefore AB blood type arises when an individual receives gene A from one parent and gene B from the other. O blood type arises when an individual receives O from both parents. A type, however, can be of two different genetic constitutions, AO and AA: the first receive A from one parent and O from the other, the second receive A from both parents. A similar situation applies to blood group B.

The existence of genetic polymorphism (a situation in which a gene exists in at least two different forms—or alleles) is demonstrated by the reaction of different blood types to specific reagents. To determine a blood type, two reagents are needed (anti-A and anti-B), which react with red blood cells (small oxygen-bearing blood cells invisible to the eye). The reaction is performed by adding two small drops of a patient's blood to a glass slide. A positive reaction occurs if, after adding a reagent, the blood cells clump together. Because blood's color is due to the red blood cells, when they clump together, the remainder of the blood becomes clear. If the reaction is negative, the blood drop remains a consistent red color. Blood group A individuals react positively only to the anti-A, while blood group B reacts only with anti-B. Those with blood group O fail to react with either serum, while AB individuals react with both.

To simplify the statistics, we do not count the number of different individuals or genotypes, but only the number of alleles—two per person. However, we have no way to distinguish between individuals of polymorphic blood group A, who could be either AA or AO. So, too, with B type blood. Luckily, simple mathematical techniques allow us to estimate how many individuals are AA and how many are AO (or BB and BO).

During World War I, Ludwik and Hanka Hirschfeld, two Polish immunologists, examined several different ethnic groups among the soldiers in the English and French colonial armies and the World War I prisoners, including Vietnamese, Senegalese, and Indians. They discovered that the proportions of individuals belonging to the different blood groups were different in every population. This phenomenon is now known to be universal. We know the number of polymorphisms is extremely high, and each human population is different for most of the other polymorphisms, as well. This early work with ABO gave birth to anthropological genetics.

Genetic Variation between Populations

The following table shows the frequency (in percent) of the ABO alleles by continent.

Region		A	B	O
Europe		27	8	65
	English	25	8	67
	Italian	20	7	73
	Basques	23	2	75
East Asia		20	19	61
Africa		18	13	69
American Natives		1.7	0.3	98
Australian Natives		22	2	76

We immediately notice wide variation among populations in different parts of the earth; each has a distinct gene frequency. The O gene always appears the majority type, varying from 61 to 98 percent. The A gene varies from 1.7 to 27 percent, while the B gene varies from 0.3 to 19 percent. If we consider smaller samples of Native Americans, the A and B genes might be completely absent.

This table suggests two questions: Is this an exceptional situation or does something similar hold true for other genes as well?

Can we explain why there is such great variation? For now, let's explore other genes and save the second question for later.

After World War I, new blood group systems were developed using the same methods that led to the discovery of the ABO system. The most complex group is the RH system, which was found among Europeans during World War II. Its study was quickly extended to several non-European populations. But aside from the ABO and RH systems, very few blood group genes have clinical importance. Anthropological curiosity—the passion to know one's ancestors, relatives, and ultimate origins—has motivated many researchers to continue the search for new genetic polymorphisms, which, performed by new genetic research techniques, is increasingly successful.

Genetics, the study of heritable differences, offers us a window through which to view that past. We know that, with few exceptions, many characteristics such as height and skin, hair, and eye color are genetically determined, but we do not understand precisely how. Moreover, some of them are also influenced by non-genetic factors, for instance, nutrition, in the case of height, and exposure to the sun, in the case of skin tone. Our poor understanding of the hereditary mechanism of these familiar characteristics is due to their interaction with non-genetic, environmental factors, and the general complexity of the mechanisms determining all traits that involve shape. By contrast, we understand clearly the inheritance of blood groups, and of chemical polymorphisms among enzymes and other proteins, because the account of traits determined by relatively simple substances like proteins is chemically simpler and easier to understand and measure. But these traits are not directly visible, and rather sensitive laboratory methods are required to detect them.

Very early on, the American scientist William Boyd showed that by using the first genetic systems discovered—ABO, RH, and MN—one could already differentiate populations from the five continents. Arthur Mourant, a British hematologist, produced the first comprehensive summary of data on human polymorphisms in 1954. The second edition of Mourant's book, appearing in 1976, contained more than one thousand pages, more than doubling the amount of data previously available.

Two major techniques are used to study polymorphisms, or genetic "markers" as they are called because they act as tags on genetic material, on proteins. One, employed for almost all blood group typings, uses biological reagents, often made by humans reacting to foreign substances from bacteria, or from other sources. These reagents are special proteins called immunoglobulins or antibodies. They are made in the course of building immunity, that is, resistance to some external agent, and usually react specifically with substances called antigens, usually other proteins. The other analytical method of genetic analysis, developed in 1948, is a direct study of physical properties of specific protein molecules, usually by measuring their mobility in an electric field. It is called electrophoresis.

Both methods revealed directly or indirectly the variation in structure of specific proteins from individual to individual. The behavior of these variants could be tested in families to confirm the genetic nature of such variation. But the number of polymorphic proteins detected in this way was small and at the beginning of the 1980s only about 250 were known. All proteins are produced by DNA, and therefore behind protein variation there must be a parallel variation of DNA, the chemical substance responsible for biological inheritance. The analytical methods necessary to chemically study DNA were developed later.

In the early eighties the analysis of variation in DNA had its start. DNA is a very long filament made of a chain combining four different nucleotides, A, C, G, and T. Changes in the sequence of nucleotides of a specific DNA happen rarely, and more or less randomly, when one nucleotide is replaced by another during replication. Thus, if a DNA segment is GCAATGGCCC, it may happen that a copy of it passed by a parent to a child is changed in the fifth nucleotide, T being replaced by C. The DNA generating the child's protein will thus be GCAACGGCCC. This is the smallest change that can happen to DNA, and is called a mutation; as DNA is inherited, descendants of the child will receive the mutated DNA. A change in DNA may cause a change in a protein, and this may cause a change visible to us.

Restriction enzymes provided a simple way to detect differences in the DNA of two individuals. Restriction enzymes are produced by bacteria and break DNA into certain sequences of 4, 6, or 8 nucleotides, for instance GCCG.

A method of multiplying DNA in a test tube with the enzyme DNA polymerase, which nature uses to duplicate DNA when cells divide, was discovered and developed in the second half of the eighties, and is called PCR, or polymerase chain reaction. This new technique has improved the power of genetic analysis in the nineties. We now know that there must exist millions of polymorphisms in DNA, and we can study them all, but the techniques for doing this at a satisfactory pace are only now beginning to be available.

The future of the analysis of genetic variation is clearly in the study of DNA, but results accumulated with the old techniques based on proteins have not lost their value. There are some specific problems, which can be resolved only by DNA techniques. On the other hand, the very rich information generated by protein data on human populations includes almost 100,000 frequencies of polymorphisms. They were studied for over 100 genes in thousands of different populations all over the earth, and many of the conclusions thus made possible and discussed in this book have arisen from studies of proteins. Results with DNA have complemented but never contradicted the protein data. We start having knowledge on thousands of DNA polymorphisms, but they are almost all limited to very few populations. We will summarize the most important ones.

Studying Many Genes Allows Use of the "Law of Large Numbers"

Is it possible to reconstruct human evolution by studying the types of living populations only? We can simplify the process of doing so by concentrating most of our studies to indigenous people, when it is possible to recognize them and differentiate them from recent immigrants to a region. But we learn much about human origins and evolution from a single gene like ABO.

We will introduce here the word "gene." Everybody has heard it, but few know its precise meaning. The old definition, "unit of inheritance," is still difficult to understand—in fact, it was used when we did not know what a gene was in chemical terms. Today we can give a much more concrete definition: a gene is a segment of DNA that has a specified, recognizable biological function (in practice, most frequently that of generating a particular protein). It is, therefore, part of a chromosome, a rod found in the nucleus of a cell that contains an extremely long DNA thread, coiled and organized in a complicated way. A cell usually has many chromosomes, and their distribution to daughter cells is made in such a way that a daughter cell receives a complete copy of the chromosomes of the mother cell. When studying evolution, however, we may, and often must, ignore what a gene is doing, because we don't know. But a gene remains useful for evolutionary studies (and others) if it is present in more than one form, and the more forms of a gene (allele) that exist, the better the gene suits our purposes. With only three alleles, ABO can hardly be very informative. In Africa, the place of origin, one finds all alleles. But this is also true of Asia and Europe. In Asia, however, the B allele is more frequent than in the other continents; group A is somewhat more common in Europe; and Native Americans are almost entirely blood group O. What conclusions can we draw? That A and B genes were probably lost in the majority of Native Americans, but why? Many have speculated about the reason, but it is impossible to provide an entirely satisfactory answer.

The first hypothesis connecting the historical origin of a people and a gene that was subsequently confirmed by independent evidence was made on the basis of the RH gene in the early forties. The simplest genetic analysis recognizes two forms: RH+ and RH−. Globally, RH+ is predominant, but RH− reaches appreciable frequencies in Europe with the Basques having the highest frequency. This suggests that the RH− form arose by mutation from the RH+ allele in western Europe and then spread, for unspecified reasons, toward Asia and Africa, never greatly diminishing the frequency of the RH+ gene. The highest frequencies of the negative type are generally found in the west and northeast of Europe. Frequencies steadily decline toward the Balkans, as if Europe was once entirely RH– (or at least predominantly so) before a group of RH+ people entered via the Balkans and diffused to the west and north, mixing with indigenous Europeans. This hypothesis would have remained uncertain if it had not been substantiated by the simultaneous study of many other genes. Archeology also lent support to the argument, as we shall see later.

Reconstructing the history of evolution has proved a daunting task. The accumulation of data on many genes in thousands of people from different populations has produced a dizzying amount of information that describes the frequency of the different forms of more than 100 genes—a body of knowledge that is very useful for testing evolutionary hypotheses. Experience has shown that we can never rely on a single gene for reconstructing human evolution. It might appear that a single system of genes like HLA, which today has hundreds of alleles, would be sufficient. The HLA genes play an important role in fighting infections and recently have become important in matching donors and recipients for tissue and organ transplants. They possess a great diversity of forms, as is necessary for a potential defense against the spread of tumors among unrelated individuals, but they are also subject to extreme natural selection related to their role in fighting infection. If the conclusions we reach about evolution through observations made using HLA are different from those obtained using other genes, we need to explain the reasons, because they may lead to different historical interpretations. It is very useful, and I think essential, to examine all existing information. The broadest synthesis has the greatest chance of answering the questions we ask, and the least chance of being contradicted by later findings.

Therefore, it is also worth gathering information from any discipline that can provide even a partial answer to our problems. Within genetics itself, we want to collect as much information about as many genes as possible, which would allow us to use the "law of large numbers" in the calculation of probabilities: random events are important in evolution, but despite their capriciousness, their behavior can be accounted for through a large number of observations.

Jacques Bernoulli, in his *Ars conjectandi* of 1713, wrote, "Even the stupidest of men, by some instinct of nature, is convinced on his own that with more observations his risk of failure is diminished."

Many studies have been invalidated because of an inadequate number of observations. When we study polymorphisms directly on DNA, there is no dearth of evidence: we can study millions. We may not need to study them all, because at a certain point additional data fail to provide new results or lead to different conclusions. Nevertheless, simply studying a large sample is not always enough. If we observe heterogeneity in our data, so that it can be divided into several categories, each implying a different history, we must further search for the source of these discrepancies. We have seen an important example in the comparison of genes transmitted by the paternal and the maternal line, as we will discuss in another chapter.

Genetic Distances

It is clear that, in order to contrast populations, we must synthesize a vast amount of genetic information. At first, to measure the "genetic distance" between populations, we simply compared pairs of populations. Only much later, when we had a very large number of genes and some new analytical techniques, were we able to study the differences among many populations, or even within individual populations. For most genes, the frequency differences between populations are nil to very slight and their contribution to the global genetic distance between populations is close to zero.

The RH gene provides interesting genetic distances in Europe, but is less useful elsewhere. For example, the frequency of RH negative individuals is 41.1 percent in England, 41.2 percent in France, 40 percent in the former Yugoslavia, and 37 percent in Bulgaria. These differences are slight, but among the Basques the frequency is 50.4 percent and among the Lapps (more appropriately called the Saami) the frequency is 18.7 percent. For this gene the genetic distance between France and England, calculated simply by taking the difference between the percentages above, is 0.1 percent. The distance between French and Bulgarians (4.2 percent) or between Bulgarians and persons from the former Yugoslavia (3 percent) is greater. But the distance between Basques and English is considerable (9.3 percent) and the difference between Basques and Lapps is dramatic (31.7 percent).

I like to explain the concept of genetic distance in the simple way that I have done above, as a difference between percentage frequencies of the form of a gene. In reality, there are now many methods for calculating genetic distances and all are fairly complicated. When I started this calculation, I asked the advice of my teacher, R. A. Fisher, one of the great geneticists and statisticians, because I could not think of a better consultant. It is pointless to give his formula here, because it is too complex. But it is still essential to average the distance between two populations over many genes if one wants reproducible conclusions.

Among other formulas subsequently proposed, one developed by Masatoshi Nei, a famous Japanese-American mathematical geneticist, has become more popular than the Fisher formula I first used. But more than twenty years after he introduced it, Professor Nei is now convinced that Fisher's approach is better than his own for the study of human populations.

In any case, most of the formulas currently used to calculate genetic distances provide very similar results overall. In fact, if I find substantial disagreement among results using the various distance measures, I tend to suspect there are other problems with the data—usually that the sample of genes is insufficient.

Once a genetic distance is calculated between populations for each of several genes, we can average all the distance values thus obtained. We thus synthesize the information from all the genes studied. The more genes we have, the more likely it is that conclusions will be correct. When we have enough genes, we can subdivide them into two or more classes and use each class to test our conclusions, which should, if everything is fine, be independent of the genes employed.

Isolation by Geographic Distance

Interesting theories developed by three mathematicians—Sewall Wright in the United States, Gustave Malécot in France, and Motoo Kimura in Japan—led, with minor differences, to the conclusion that the genetic distance between two populations generally increases in direct correlation with geographic distance separating them. This expectation derives from the observation that while most spouses are selected from within their own village or town, or part of a city, a small proportion are chosen from neighboring ones. This proportion reflects the migration that goes on all the time everywhere because of marriage. In the simplest model, equal numbers of migrants are exchanged between neighboring villages. The first measurements of migration arising from marriage were performed by Jean Sutter and Tran Ngoc Toan, and independently by myself in collaboration with Antonio Moroni and Gianna Zei, using church wedding records, which noted the spouses' birthplaces. They confirmed the tendency of people to find spouses from a short distance away, as expected. The first verification of the theory that genetic distance increases with geographic distance between populations was provided by Newton Morton, who studied small, homogeneous regions. Menozzi, Piazza, and I extended them to the entire world in our book *The History and Geography of Human Genes,* from which figure 1 was taken.

The increase of genetic distance with geographic distance may be linear at first, but over a greater geographic distance, the increase in genetic distance slows sharply. The two characteristics of the curve—the rate (i.e., the slope) of the initial increase, and the maximal value reached by the genetic distance over a great geographic distance—are different for the various continents. They are greatest for indigenous Americans and Australians, and slightest in Europe, which is the most homogeneous continent. The maximal genetic distance (in Europe) is three times smaller than on the least homogeneous continents. Despite political fragmentation, migration within Europe has been sufficient to create a greater genetic homogeneity than elsewhere. The curve has not reached a maximum value (and therefore

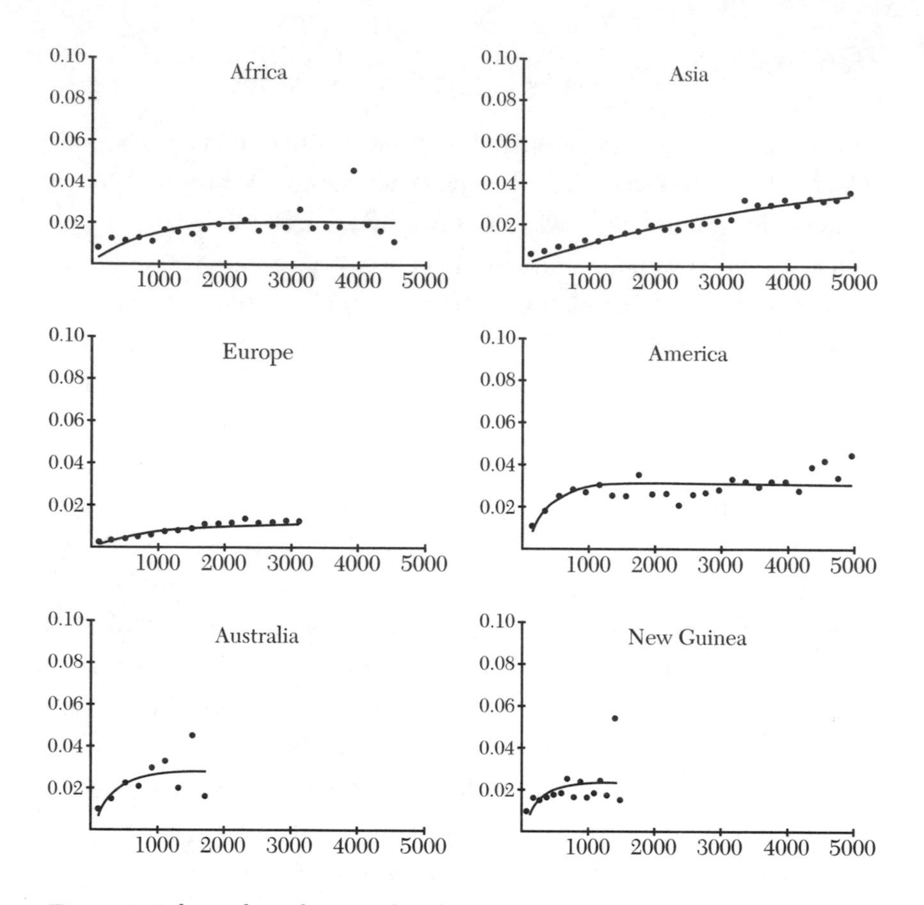

Figure 1. Relationship of geographic distance (in miles, on the horizontal axis) to genetic distance (on a scale between 0 and 1, vertical axis) in the various continents. Genetic distances among pairs of populations were averaged for all available data on 110 genes tested by methods of protein analysis (blood groups, electrophoresis, etc.). Robust averages of genetic distances were calculated for all possible pairs of tribes, towns, or other human communities that share a geographic distance class. (Cavalli-Sforza, Menozzi, and Piazza 1994)

the point of genetic equilibrium) in Asia (and clearly even less in the whole world) in spite of the extensive migrations of the past millennia. Mongols, for example, began important expansions east, south, and west around 300 B.C. The Turkish advance, halted near Vienna in the eighteenth century, was their last exploit.

Figure 1 shows the remarkable precision with which the data support the theory. Naturally, individual pairs of populations would vary substantially from the theoretical curve, but the points in figure 1 are the averages of many population pairs, calculated from over a hundred genes. We observed that it matters little which genes are chosen. Only one genetic system shows a major deviation from the others—the immunoglobulin genes. These genes code for our antibodies, and the greater variation found in them is probably a response to the great geographic variation in the array of infectious diseases we encounter.

What Is a Race, Then?

A race is a group of individuals that we can recognize as biologically different from others. To be scientifically "recognized," the differences between a population that we would like to call a race and neighboring populations must be statistically significant according to some defined criteria. The threshold of statistical significance is arbitrary. The probability of reaching significance for a given distance increases steadily with the number of individuals and genes tested.

Our experiments have shown that even neighboring populations (villages or towns) can often be quite different from each other. There is a limit to the number of individuals in a given village population who can be tested. But the maximum number of testable genes is so high that we could in principle detect, and prove to be statistically significant, a difference between any two populations however close geographically or genetically. If we look at enough genes, the genetic distance between Ithaca and Albany in New York or Pisa and Florence in Italy is most likely to be significant, and therefore scientifically proven. The inhabitants of Ithaca and Albany might be disappointed to discover that they belong to separate races. People in Pisa and Florence might be pleased that science had validated their ancient mutual distrust by demonstrating their genetic differences. In his *Divine Comedy,* Dante, a Florentine, expressed his dislike of people from Pisa by wishing that God would move two islands situated at the mouth of the river Arno, thereby flooding Pisa and drowning all its people.

Classifying the world's population into several hundreds of thousands or a million different races would, of course, be completely impractical. But what level of genetic divergence would be necessary to determine boundaries for a definition of racial difference? Because genetic divergence increases in a continuous manner, it is obvious that any definition or threshold would be completely arbitrary.

It has been suggested that one might define race by the analysis of discontinuities in the surface of gene frequencies generated on a geographic map. Introduced by Guido Barbujani and Robert Sokal (1990), the method looks for local increases in the rate of change of gene frequencies, per unit of geographic distance. Obstacles to migration or marriage could create these local increases. If proved for many genes, such barriers could help distinguish races. But a true discontinuity is difficult if not impossible to establish for gene frequencies, so they would rather look for regions where gene frequencies change rapidly. The particular rapidity of genetic change that could suffice as a "genetic barrier" would naturally be chosen in an arbitrary manner.

This procedure illustrates the theoretical difficulties classification by race poses. Gene frequencies are not geographic features like altitude or compass direction, which can be measured precisely at any point on the earth's surface; rather, they are properties of a population that occupies an area of finite extent. One possible solution would be to use villages and small cities as "points" in geographic space. Large cities could be subdivided into several points to take account of residential segregation. But the available data on gene frequency in villages or small cities are insufficient and they would provide an extremely detailed clustering.

In any case, this method is still useful for identifying the geographic location of genetic "boundaries," however arbitrary these are. In Europe, for example, Barbujani and Sokal found 33 genetic boundaries that corresponded in 22 cases to geographic features (mountains, rivers, seas) and almost always (in 31 cases) to linguistic or dialectic boundaries. In a country with a homogeneous language, like Italy, family names provided better results than genes. Because they're inherited, surnames can give almost the same information as genes, but are more informative because surnames are readily available in large numbers.

A more significant difficulty resulting from racial classification is that the barriers found by the method described above have rarely defined a closed space inhabited by a population enclave, even when aided by geographical features such as the Alps. Islands may be the only exceptions. The population of each island could be classified as a race, because it would be different from other islands and the nearest mainland, if there were sufficient genetic information. But would that be useful for practical purposes, like for instance taking a census in the United States? The answer is certainly no. A third problem is that a huge number of genes must be studied to distinguish closely related populations.

Scientific attempts to classify races continued through the end of the nineteenth century. The results often contradicted each other, a good indication of the difficulty of such efforts. Darwin understood that geographic continuity would frustrate any attempt at classifying human races. He noted a phenomenon that repeated itself many times in the course of history: different anthropologists come to completely different tallies of races, from 3 to over 100. But why does this compulsion to classify human races exist? The question is extremely important. Maybe it would be more useful to answer a more general question: why classify?

Why Classify Things?

When we are presented with a great number of things, we feel compelled to impose some order on potential chaos. Such is the goal of classification. It allows us to describe a complex array of objects with simple words or concepts, even at the cost of oversimplification.

Zoologists and botanists have classified thousands or even millions of species, and their work is not close to being finished. If variation were not important and complex, it would not be necessary to categorize at all. One could simply recognize the level of difference relevant to one's needs.

Humans are not alone in their tendency to classify. Chimpanzees, for example, and probably most other animals, can separate several hundred leaves and fruits into edible and non-edible categories. Depending on their appetite, other categories may be used, although edibility is fundamental since many plants are potentially toxic. Chimpanzees have even been observed teaching their offspring which foods can be eaten and which cannot.

Unlike animals, humans use language to differentiate between objects. We assign a name to each object we wish to distinguish. African Pygmies recognize hundreds of tree species (Western botanists identify a similar number) and several hundred animals; but such diversity is still too little to require a terribly high order of classification.

Classification and some accompanying oversimplification become necessary when variation is very high. Naturalists such as Georges Louis Leclerc Buffon and Carolus Linnaeus established valid systems of classification for the extraordinary diversity of plant and animal species. Similar systems can be found in some so-called "primitive" populations who have an undeveloped (or non-monetary) economy.

Why can classifying human races be useful? Demographers and sociologists undoubtedly have some opinion on the subject. Most practical classifications are extremely simplistic. The U.S. census recognizes Whites, Blacks (African Americans), Native Americans, Asians, and Hispanics. This last category has almost no biological meaning. In practice it refers to Mexicans, but more generally, a large number of Spanish-speaking people are assigned to it.

Proposing an improved classification can only end in failure. Observing the variation between ethnic groups should convince us of that. Visible differences lead us to believe in the existence of "pure" races, but we have seen that these are very narrow, essentially incorrect criteria. And when measured and plotted carefully, visible traits are actually far less discontinuous than is usually believed. Classification based on continental origin could furnish a first approximation of racial division, until we realize that Asia and even Africa and the Americas are very heterogeneous. Even in Europe, where the population is much more homogeneous, several subdivisions have been proposed. But it is immediately clear that all systems lack clear and satisfactory criteria for classifying. The more we pay attention to questions of statistical adequacy, the more hopeless the effort becomes. It is true that strictly inherited characteristics are more satisfactory than anthropometric measurements or observations of colors and morphology. But above all it is true that one encounters near total genetic continuity between all regions while attempting to select even the most homogeneous races.

The observation has been made that almost any human group—from a village in the Pyrenees or the Alps, to a Pygmy camp in Africa—displays almost the same average distance between individuals, although gene frequencies typically differ from village to village by some small amount. Any small village typically contains about the same amount of genetic variation as another village located on any other continent. Each population is a microcosm that recapitulates the entire human macrocosm even if the precise genetic compositions vary slightly. Naturally, a small village in the Alps, or a Pygmy camp of 30 people, is somewhat less heterogeneous genetically than a large country, for example, China, but perhaps only by a factor of two. On average, these populations have a heterogeneity among individuals only slightly less than that in evidence in the whole world. Regardless of the type of genetic markers used (selected from a very wide range), the variation between two random individuals within any one population is 85 percent as large as that between two individuals randomly selected from the world's population.

It seems wise to me, therefore, to abandon any attempt at racial classification along the traditional lines. There is, however, one practical reason for keeping an interest in genetic differences.

Can the Study of Genetic Differences Be Useful in Practice?

The intellectual interest of a rational classification of races clashes with the absurdity of imposing an artificial discontinuity on a phenomenon that is very nearly continuous. But is there a practical reason that justifies it? It clearly must be sought where a real discontinuity exists, if any. Here, it happens, we are closer to practical reasons that justify some sort of classification on the basis of genetic differences.

Humans live in social communities. The social group is evolving rapidly, its size increases and its internal structure becomes more complicated. The majority of the world is still, however, formed by groups that are at the lower end of the complexity scale. The industrialized countries are at the opposite end. Most people like to identify with their social group, and therefore give it a name. For obvious reasons, this tends to be the same as the name of the language, and of the tribe, although in many cases the tribe has grown enough that it is no more a simple social group. Within the larger groups there tend to be further subdivisions. This helps to give a lower bound of the number of social human groups existing on Earth. The number of languages existing today is 5,000–6,000, and the number of social groups that may exist today in the world must be greater than 10,000, or even 100,000.

If we want to fix an upper bound we must be more precise about the meaning of social group. From a genetic point of view, the most meaningful social group is the one in which one is likely to find a spouse. The minimum size of such a group to avoid deleterious effects of inbreeding is five hundred. This is also a "magic number" that many anthropologists, not without some factual support, indicate as the average size of a tribe, especially for the more economically primitive ones. This would mean that there exist on Earth at most ten million social groups. On the basis of some other considerations, perhaps one million may be a reasonable upper bound of the number of social groups that are worth being considered as distinct from a genetic point of view. The average group would consist of 5,000–500,000 individuals. These numbers may have to be somewhat modified and I reserve the right to do it. But the principle remains valid.

Clearly no anthropologist would accept a classification into a million races, and probably not even into 10,000 ones. But this is a "genetic" classification that might be useful, and will probably exist, with some further complication, a day not too far ahead. Individuals belonging to a group of this kind would have genetic similarity greater than two random individuals, because they would share significantly more ancestry. In fact, the group will have been defined on the basis of endogamous behavior (a tendency to marry within the group). Endogamy tends to generate, gradually, both genetic and cultural differentiation between groups. We have seen that the genetic differentiation of populations, even if real, is small, but stable in time. In contrast to this, cultural differentiation can be surprisingly high, and fast to reach, but also more easily reversible, and hence less stable. But there is no question that genetic differences can be important from a very practical point of view: namely the chances of having specific diseases, and responding similarly to the same drugs.

To skeptical readers, an example of the application of this principle can already be seen in Iceland, where medical research of all Icelanders has begun with Parliamentary approval by a foreign pharmaceutical firm. Here the population is of 250,000 individuals, and therefore within the upper and lower bounds defined before. But current research may show that the Icelandic population is not as homogeneous as might be expected.

Weakness and Strength of Historical Research

We have begun a survey of human diversity, and it is inevitable to ask oneself a number of questions: How are such diversities produced? What are the forces at work? What has been the course of these events? In short, what has been the history of human evolution, and which factors have caused and directed it?

Any attempt at reconstructing human evolution presents the same problems we encounter in historical research. Experimental science allows us to test any hypothesis, no matter how unlikely, but history cannot be repeated at will—even if it sometimes gives the appearance of repeating itself. Nevertheless, historical and anthropological analogies are often useful. When these offer independent confirmation or supplementary evidence, they allow us to eliminate or support a hypothesis. Multidisciplinary research provides, in a way, a sort of replication of an event, which is generally possible only in experimental science.

Exploring related disciplines can lead to rich discoveries. It was with this intent that I have searched for, and often found, support from fields such as linguistics, archeology, and demography. Just as this approach yields positive results, it is also a source of great intellectual satisfaction. The researcher sees the fundamental unity of the sciences and their procedures.

CHAPTER 2

A Walk in the Woods

Several years ago, I found myself wondering whether it was possible to reconstruct the history of human evolution using genetic data from living populations. At that time, our knowledge about the subject came mainly from paleoanthropology. But fossil material was scant, and even today we must content ourselves with a very small number of incomplete skulls and bones. Those few fragments are the only random pieces left of a giant jigsaw puzzle—how can we hope to reconstruct the whole from such limited clues? Often a new fossil, or the revision of a single date, forces a major reassessment of our understanding of human evolution—the discovery of a million-year-old mandible may take up entire pages in the scientific and popular presses.

Our obsession with fossils has distracted us from a much richer source of evolutionary information: genetic data, although largely restricted to living populations, can tell us a great deal about human history. Genes and gene frequencies, unlike skeletal characteristics, change over time according to precise and well-understood rules.

Of course skeletal morphology, or the evolution of bones, is also genetically determined, but in a much more complicated and less clearly understood manner. It is influenced by many factors, most notably the environment.

Genetics, too, has its shortcomings. It is difficult, for example, to study ancient populations. But we now know that DNA is occasionally preserved in fossils that are not too old. A new science of paleogenetics is emerging, as any reader of *Jurassic Park* would know. Claims have been made that DNA can sometimes be recovered from insects preserved in amber for several million years. Of course elephants or humans won't show up in amber deposits. Moreover, it seems that research on DNA found in amber was overly optimistic, and that there is little, if any, hope of finding DNA in good condition from organisms that lived millions of years ago.

The problem is that old DNA undergoes extensive fragmentation and chemical alteration, and it is only by comparing many fragments of the same DNA segment that one can confidently reconstruct the complete structure of even a single short segment. Such a procedure was successfully attempted in the Munich laboratory directed by Svante Paabo, a student of Allan Wilson, a pioneer on studies of mitochondrial DNA. The first samples studied successfully came from a skeleton nicknamed Oetzi, a Bronze Age man whose body was uncovered by melting ice in the Alps between Italy and Austria. Oetzi's clothes and tools have given us precious information about Bronze Age fashion and technology. The DNA studied in Oetzi's remains comes from very small bacteria-like bodies called mitochondria found in the cells of every higher organism from yeasts to mammals. Most of our cells usually contain hundreds or thousands of mitochondria. Each cell has at least one copy of a little chromosome made of DNA. Thus, a considerable amount of mitochondrial DNA (mtDNA) is present in almost every cell. There is a good reason: mitochondria are needed to generate energy for cell growth and maintenance, using chemical nutrients. By contrast, DNA forming chromosomal genes is present in each cell in only two copies per gene: one from the father, the other from the mother. Thus it is more likely that a sufficient amount of mtDNA will be preserved.

Oetzi's mtDNA was strikingly similar to that found in people living in the same general region today. The population of the area must have been reasonably stable, with little migration of outsiders occurring during the 5,000 years since Oetzi's death.

The tour de force paid off when a much more ambitious task was attempted in the same laboratory—the extraction of DNA from a Neandertal specimen.

In 1856, excavation work in northern Germany uncovered a skull that clearly differed from those of modern humans. The idea of evolution was still vague—Darwin's *The Origin of Species* would not be published for another three years. However, a local schoolmaster who heard about the discovery understood the significance of the find and took the skull to the professor of anatomy at Bonn University. The skull was named after the place of discovery, Neander Thal (the valley of the Neander River).

Many other similar fossils have been found in the one hundred fifty years since then but the relation between this man and modern humans still vexes physical anthropologists. The differences are clear but there is also a remarkable sense of familiarity. Some thought Neandertals (also spelled with an "h") were the direct ancestors of modern humans. Others thought they were an extinct branch of an older human type. This question could be resolved by analyzing the DNA of the extinct Neandertals, if any could be found. A bone sample from an ancient humerus was examined in the same Munich laboratory with an approach similar to that used to study Oetzi's remains. The result was unequivocal. There is a considerable difference between the mtDNA of this Neandertal and that of practically any modern human. From a quantitative evaluation of this difference it was estimated that the last common ancestor of Neandertal and modern humans lived about half a million years ago. It is not quite clear where common ancestors lived, but modern humans and Neandertal must have separated early and developed separately, modern humans in Africa and Neandertals in Europe. The results of mitochondrial DNA show clearly that Neandertal was not our direct ancestor, unlike earlier hypotheses made by some paleoanthropologists. Around 60,000 years ago Neandertals expanded from Europe to Central Asia and the Middle East, but there are no later finds of them in these regions. Modern humans arrived in Europe 42,000–43,000 years ago; they may have had contacts with Neandertals, but no evidence of hybrids was found. Shortly after 40,000 years ago Neandertals become more and more rare in Europe, and the last specimens found so far are about 30,000 years old.

These painstaking studies will hopefully be extended to many other Neandertals and other old skulls, and in the future they may help clarify recent hominid evolution. Unfortunately the method is not easily effective with older samples, and studies on nuclear genes, which would be more informative, have been successful only on fossils of Oetzi's age or younger.

My own interest in human evolution started at Cambridge University in the Genetics Department chaired by R. A. Fisher. The first ten years of my research were in bacterial genetics, but in 1951, when I began teaching general genetics at the University of Parma in Italy, I shifted my attention to humans—a more charismatic organism. By 1961 I felt there were sufficient data to tackle the problem with which this chapter is concerned.

Growing an Evolutionary Tree

Ever since Darwin, we have thought of evolution in terms of trees that trace the relationships among species and their ancestors. By definition, a species is a group of individuals capable of mating and producing fertile offspring. Humans comprise a single species, and all populations are interfertile. This implies that all human groups share a recent common ancestry, and/or that genes may continuously be exchanged between populations. On the other hand, if populations split relatively cleanly, with little or no subsequent genetic exchange between the two new sub-populations, a tree is an accurate representation of the process. When a new continent is occupied, a discrete fission usually results. Migration from one continent to another necessarily involves the development of some discontinuity. Even if the migration takes much time and there remains geographic contiguity between mother and daughter populations, some degree of genetic differentiation is the ultimate result.

As I've said in chapter 1, it is easy to determine the genetic relatedness of several populations by calculating genetic distances between pairs of them. Let us take the modern indigenous populations of the five continents—a simpler proposition than the fifteen populations (three per continent) that Anthony Edwards and I first studied in 1961. The genetic distances between continents, expressed as percentages, are as follows:

Genetic Distances between Continents

	Africa	Oceania	America	Europe
Oceania	24.7			
America	22.6	14.6		
Europe	16.6	13.5	9.5	
Asia	20.6	10.0	8.9	9.7

Starting from these genetic distances, how can we construct a tree that illustrates the successive fissions that have produced these differences?

The methods Edwards and I developed originally for this task were rather complicated, but for the sake of illustration we will choose a simple technique called average linkage. We later learned that average linkage actually produces almost the same results as more reliable, but much more difficult, methods.

Easy analysis of the five continents by a tree is assured when the groups are already arranged in a rational order, as in the table "Genetic Distances between Continents." We first look for the smallest distance—the one between Native Americans and Asians. It is reasonable to expect that the longer the time of separation between two populations, the greater the genetic distance between them; the separation between Asia and America should therefore be the most recent fission of the tree. In fact, we know from archeological information that the Americas were probably settled between 10,000 and 25,000 years ago, when a land bridge connected Siberia and Alaska during the last ice age, making it possible to walk from Asia to the Americas. There are still uncertainties about the settlement dates, as we will discuss later in more detail, but it is probable that America was the last continent occupied by modern humans.

The proportionality between genetic distance and time of separation is a reasonable principle but is not always necessarily true. The distance between America and Asia is the smallest, but it is smaller than those between Europe and America and between Europe and Asia by a very small margin. All measurements are affected by statistical error, and therefore the "true value" of the quantity measured is never really known. We can never reach perfection in measurements, but we can estimate the statistical error, and decrease it at will by increasing the number of observations. The distances we are using are based on about one hundred genes, and yet they are affected by a statistical error around 20 percent. This quantity gives us a way to calculate an interval within which the true value may lie, with a given probability. And we can always reduce the error if we can increase the number of genes. It may be enough to say for now that we have other, more complicated data that reinforce our statement. Let us therefore accept that the smallest distance is that between Native Americans and Asians and therefore the most recent split is between Asia and America. We begin our tree thus:

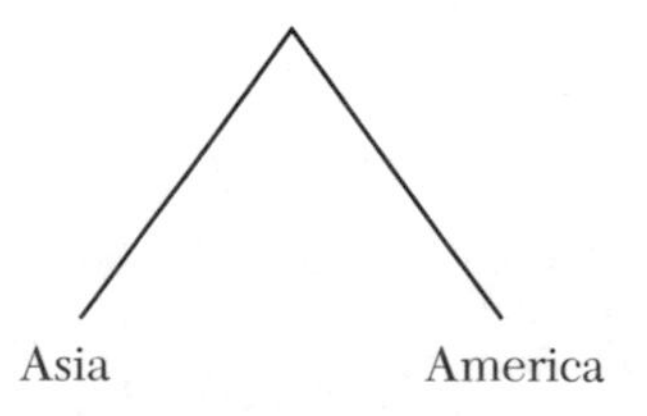

Now we can combine the two continents by calculating the mean between the distances of each of the other continents with Asia and America (for example, the genetic distance between

Europe and Asia is 9.7 while that between Europe and America is 9.5; their mean is 9.6). The preceding table loses one line and one column and becomes:

	Africa	Oceania	Asia-America
Oceania	24.7		
Asia-America	21.6	12.3	
Europe	16.6	13.5	9.6

We again select the shortest distance. This time it is between Asia-America and Europe. We add a new branch to the tree by join ing Asia-America to Europe:

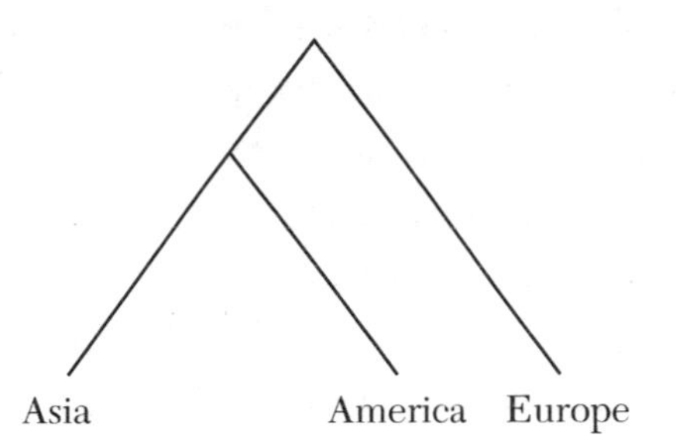

Repeating the procedure, we add Oceania, and at the final step, Africa. The final tree is:

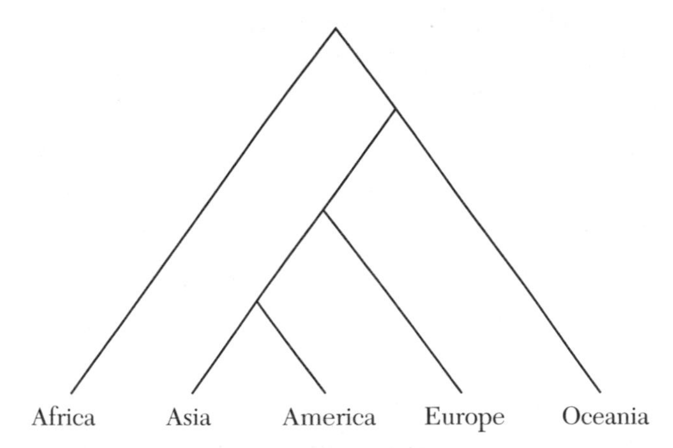

A model of human migration from Africa—arriving first in Australia, then in East Asia, and finally in Europe and America—is eventually borne out by these distances and with the archeological data. This tree, therefore, has a reasonable probability of accurately representing the evolution of modern humans. We will see later how it fits the settlement dates.

In this tree, I have simplified our task by eliminating the most obviously admixed populations of North Africa, western Asia, and the Pacific islands smaller than Australia (but combining New Guinea with Oceania because New Guineans resemble Aboriginal Australians—also included in Oceania). Nearly a fourth of the world's populations have been ignored in our example. But even if I hadn't made these cuts, similar results would be obtained, since the tree-building method has proven more reliable than we initially thought.

Nevertheless, tree building methods can fail too. One reason is that human populations constitute a genetic continuum. When we divide a continuum, only very arbitrary results can be obtained. Darwin recognized this and denounced attempts at classifying races. The potential for statistical error while calculating genetic distance is enormous and the only way to conquer this powerful impediment is to increase the number of genes analyzed, which of course requires much more work. When we first started gathering published data of gene frequencies in human populations in 1961, only twenty genetic *polymorphisms* were known for the fifteen populations, three per continent, which we wanted to employ, but by 1988 we had nearly a hundred, which has reduced the statistical error by a factor of more than two. Today, several hundred genes are known and statistical error has been reduced further. Results remain similar, but there are other methodological sources of uncertainty beyond statistical error. Fortunately we do not have to wait for another thirty years to pass to respond to them, since recently developed approaches can help us understand the new problems that have arisen.

An Enormous Forest

Different tree building methods can give slightly different results for the same group of observed values. A practical limitation to the search for an accurate tree is the extremely large number of potential trees for a given set of populations (also called, more generally, taxa, the plural of taxon). With three taxa, A, B, and C, we must choose between three trees to locate the root, R:

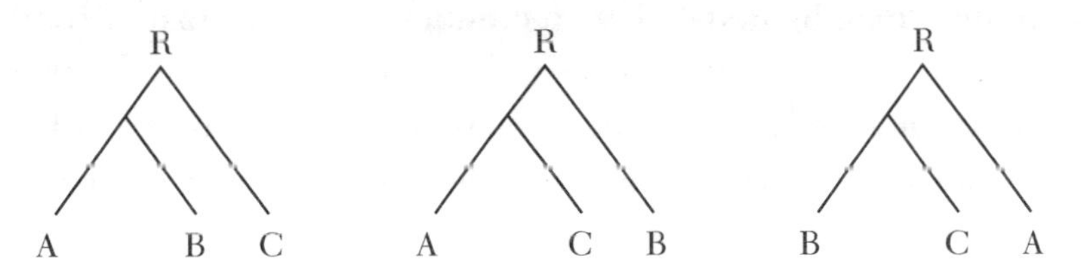

With four populations, there are 15 different rooted trees. But ignoring the root, there remain only three possible trees:

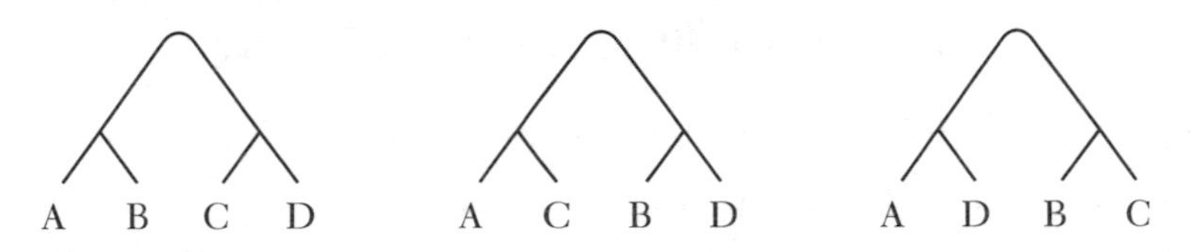

The number of possible trees grows very rapidly when one increases the number of studied populations. With five populations, there are 105 rooted trees and 15 unrooted trees. With ten populations there are 34,459,425 rooted trees and 2,027,025 unrooted trees; with twenty populations the number of possible trees is approximately 8×10^{20} and 37 times fewer without the root. In theory it would be necessary to analyze all the possible trees to be sure of finding the best one.

The situation is worse if one wants to use a more appropriate method offered by modern statistics—maximum likelihood. With this method, the calculations take an even longer time, but the advantage is that it permits a truly rigorous approach to the analysis of data. Using this procedure, one must first create a precisely defined evolutionary hypothesis (or "model") and use the observed data to test it. It is possible to obtain a measure of the "goodness of fit" with which the chosen model represents the data. If we want to test multiple hypotheses, the method allows us to choose the most satisfactory one, and evaluate its advantage over the next best tree. Unfortunately, today's computers cannot exhaustively examine all the possible trees for greater than a dozen or so populations with the more advanced methods.

The simplest maximum-likelihood model assumes identical evolutionary rates in each population, producing a rooted tree with all branch lengths equal. This technique is approximated by the method of average linkage, which also assumes equal evolutionary rates. But are evolutionary rates always the same? To find out, we must examine the factors that affect the pace of genetic change.

Evolutionary Mechanisms and Rates: Survival of the Fittest and of the Luckiest

From the beginning of modern genetics, four evolutionary forces have been recognized: mutation, which produces new genetic types; natural selection, the mechanism which automatically selects the mutated types best adapted to a particular environment; genetic drift, the random fluctuation of gene frequencies in populations; and migration, sometimes called gene flow.

Genetic drift is the most abstract of these forces, but it is really nothing more than the chance fluctuation of gene frequencies over several generations. American Indians have an O blood type frequency nearing 100 percent, and yet they probably descend from an Asian population with a group O frequency around 50 percent. If, as some have supposed, the number of immigrants from Asia to America was on the order of a dozen or less, they could all have been type O. The probability of ten Asian immigrants having this same blood type is around one in a thousand—small but not entirely negligible. If they were five, the probability would be one in thirty-two. If the first colonists were all group O, all their descendants would have been group O as well, unless new mutation or later migrations introduced other blood groups. This extreme example illustrates a statistical property of population size: the smaller a population, the greater the fluctuation in the relative gene frequency over the generations. This extreme form of drift (also called, for greater precision, "random genetic drift" to differentiate it from other meanings of the word "drift") is also called "founders' effect," but it does not have to occur specifically, or only at the founding event. It occurs at every generation, being more marked the smaller the population size. Founders' effect is likely to be of major importance because founders are usually few, and if a new settlement is successful, the population tends to increase rapidly after the beginning.

A remarkable property of drift is its tendency to homogenize populations. If genetic drift operates without introduction of genes by migration or mutation, a population will eventually eliminate all but one genetic type. If two populations with identical gene frequencies separate, both will eventually become fixed for a single genetic type, but possibly different in the two populations. Drift acts blindly—the increase or decrease of gene frequencies of a given type in a drifting population is due entirely to chance. As a result, the behavior of drift can only be predicted in terms of probabilities. For example, the most common gene at the beginning of the process is the one most likely to reach final fixation in the population. This probabilistic nature of drift prompted Japanese population geneticist Motoo Kimura to modify the Darwinian stereotype of evolution, "survival of the fittest," to incorporate also the role of chance in evolutionary change—what he called "survival of the luckiest."

Is the lack of the A and B blood groups in the Americas due to drift? We cannot be sure, but we should also consider an alternative hypothesis—natural selection. Infectious diseases have been a major cause of human mortality. Some genes, including the ABO blood group genes, can convey resistance to particular infections, and there is some evidence that the O blood group imparts resistance to syphilis. According to one popular hypothesis, syphilis was common in the Americas and was brought back to Spain in 1493, believed to be its first appearance in Europe, by the first Spanish sailors of Columbus's expedition. Hastened by war, it spread to France and Italy and soon to the rest of Europe. Research on pre-Columbian mummies suggested that the A and B blood groups existed in the Americas several thousand years ago, but this has not been substantiated by modern analytical methods. If the result could be confirmed, however, it would implicate natural selection in the disappearance of the A and B genes from the Americas. If the O blood group confers some resistance to syphilis, and there are some cues that it may, its frequency would increase relative to the susceptible A and B blood groups during an epidemic.

The choice between these two hypotheses—drift and natural selection—is often difficult to make. One must understand that natural selection is simply a demographic phenomenon: certain genetic types within a population may have greater or smaller probability of yielding children than has the rest of the population. This may happen because they have a greater, or smaller, probability of resistance to some adverse condition, for instance, an infectious disease. Or they may have greater or lower fertility. Natural selection affects evolution of the traits responsible for the difference of mortality or fertility only if the traits are transmitted to progeny, as is typical of genes. Thus natural selection usually affects a particular gene, favoring certain forms (alleles) of the gene, which will find themselves at an advantage over others (but usually only in specific environments). Thus the three ABO alleles confer different resistance to diseases like *E. coli* infections, tuberculosis, perhaps syphilis and smallpox. O individuals are more susceptible to gastroduodenal ulcer, which we have recently learned is determined by the bacterium *Helicobacter pylori*.

While natural selection affects each gene in a very specific way, genetic drift affects all genes according to the same, clearcut probability laws. Under drift, the average magnitude of change of frequencies is the same for all genes and is determined by the size of the population from generation to generation. However, whereas the direction of change is random for each gene under genetic drift, natural selection acts only on some genes and determines change in a particular direction. In very large populations, natural selection is relatively unimpeded by drift. In the special cases of small isolated populations, such as those living in the mountains or on islands, drift is more likely to be a dominant force, though natural selection is the real creative force of nature.

We sometimes wonder how natural selection could have created the magnificent organs and functions of living organisms, like the eye or the ear. It may seem extremely unlikely that such perfect and complex organs ever developed, but natural selection is the force that can create improbability, because it picks out automatically very rare novelties produced by mutation, anytime they carry an advantage for the organism in its specific environment. Of course, organs as complicated as the eye or the ear are not created in one generation or by one mutation, but by the accumulation of very many changes that have operated in the same directions.

Natural selection can target any gene. Because mutations are random changes in genes that have been adapted for specific, intricate functions over many millions of years, they are frequently deleterious, causing sickness or death. Natural selection will automatically eliminate mutations that lower the survival or reproductive output of those who carry them. Nevertheless, many genetic changes are neither beneficial nor disadvantageous; they are "selectively neutral" and have the greatest chance of experiencing random drift. In the absence of historical data, it is difficult to distinguish between diffusion of a selectively neutral gene into a population because of drift and the spread of an advantageous mutation by natural selection. In some few cases selection can work quickly, but the selective advantages or disadvantages of specific alleles are usually modest, and it may take thousands or even tens of thousands of generations to substitute an improved gene. In humans, one thousand generations would span about 25,000 years. If a gene presents a strong selective advantage, it can be spread by natural selection in only a few hundreds, or few thousands, of years. This almost certainly was the case in northern Europe and parts of Africa as adults grew capable of digesting lactose, the sugar found in milk. Children everywhere can digest milk until the age of three or four, but they generally lose this ability when they cease drinking their mother's milk. In those populations that herd sheep, goats, cattle, and other animals, adults have frequently started drinking milk. There is a strong selective advantage to digesting lactose into adulthood. Animals were domesticated only within the last 10,000 years and yet, within that time, the capacity to digest lactose has reached almost 100 percent in herding populations in which adults drink milk.

So evolution by natural selection can, in fact, be particularly rapid for genes that impart a significant selective advantage to their bearers. Selection for an advantageous form of a gene will usually result in the fixation of genes that began at a very low frequency in the population. Every evolutionary process begins with a mutation appearing in a single individual. Even when a mutation is strongly advantageous and increases in number in successive generations, it is represented by only a few individuals (usually one) at first. It is therefore subject to drift, which may even eliminate a potentially successful mutation. As inheritors become more numerous, random extinction becomes unlikely.

In summary, natural selection is determined by the difference between the mortality and/or fecundity of different genetic types (also called "genotypes"). Those genes that reduce mortality or increase fertility will increase in frequency in subsequent generations. The genotypes that increase mortality, especially among the young, or that reduce reproductive output tend to disappear from the population. The biological adaptation of an individual to the environment in which he lives is measured solely by his capacity to survive and to reproduce. The process is completely automatic, and the "survival of the fittest," or more accurately, the greater representation in future generations of those who have better chances of surviving and reproducing (i.e., are genetically fitter), is the cornerstone of natural selection.

Heterozygote Advantage

During the nineteenth century, the concept of racial purity received much attention. The perfection of races or breeds is still an important goal for animal breeders. Dog and cat shows establish, often arbitrarily, an ideal of esthetic perfection, which trainers wish to attain with their animals. This is frequently a counterproductive effort. Breeders know that by seeking genetic purity through repeated crossings between closely related animals—inbreeding—they dangerously reduce the animals' resistance to disease. The reverse—outcrossing—is more desirable since racial mixing in all species generally increases disease resistance and overall viability. This phenomenon is known as "hybrid vigor." When considering the hybridization of a single gene, one speaks of heterozygote advantage. A heterozygote is an individual who receives different forms of a gene from father and mother.

The classic example of heterozygous advantage is sickle cell anemia, which affects mostly, but not exclusively, Africans. Consider a parallel example, common in people of southern European origin: a gene responsible for a genetic disease called thalassemia, a severe anemia that usually kills before reproductive age is reached. The gene shows up in two slightly different forms, or alleles: normal N, and abnormal T (causing thalassemia). There are three possible genetic types:

NN: individuals who receive the normal gene N from both parents are *"normal" homozygotes.*

NT: those who get a normal allele N from one parent and a thalassemia gene T from the other are *heterozygotes.* Like normal homozygotes, they do not have the disease (but can be identified through simple laboratory blood tests).

TT: persons who receive a thalassemia allele from both parents are *homozygotes for T,* the abnormal gene, and have the disease.

In some European populations, for instance in the Italian province of Ferrara, located between Venice and Bologna, one of approximately one hundred children is born with thalassemia. Nearly all those afflicted die young. Heterozygotes are 18 percent of the population and the rest, 81 percent, are normal homozygotes.

The important question is: why do so many people have the disease, since they inevitably die before they reach adulthood? They are obviously at a selective disadvantage, and the disease should disappear through natural selection. The reality is, however, more complicated; the province of Ferrara has also been affected for many centuries by a highly lethal infectious disease, malaria. It so happens that heterozygotes for thalassemia are resistant to it, although normal homozygotes frequently succumb to the infections. The incidence of malaria in the Ferrara region was so high until World War II, that about one out of ten normal homozygotes died from it, while heterozygotes almost always survived it. Given these numbers and a few calculations we might see that an equal proportion of N genes and of T genes disappears at every generation, the first because of malaria and the second because of thalassemia. Therefore, until there is malaria of sufficient strength, thalassemia remains at a stable frequency in the population. The thalassemia allele gives the population some protection from malaria: in fact, it saves 8.1 percent of the N homozygotes who would otherwise die because of malaria, at the cost of a smaller number of deaths (1 percent) because of thalassemia.

If malaria disappears, thalassemia will also disappear, because normal homozygotes and heterozygotes will survive at the same rate, while homozygous TT will continue to die young. As malaria becomes more or less intense, the frequency of the thalassemia gene will increase or decrease.

We have said this type of natural selection is called heterozygous advantage. Whenever the two homozygotes survive or reproduce less than the heterozygote, the two alleles will remain in the population and adjust their frequencies automatically to values such that equal proportions of the two genes survive every generation. At this point, which may be reached after a relatively few generations have passed, the relative frequencies of the two alleles, and therefore of the three genotypes, do not change any further.

Sickle cell anemia is a disease that shares many features with thalassemia, and is especially frequent in people of African, Arabic, and Asian Indian origin. In the case of genes like those producing thalassemia, of which there are many different types, or sickle cell anemia, the mutant has no chance of reaching a high frequency, since its incidence is governed by a delicate equilibrium between the selective advantage of the heterozygotes and the usually serious disadvantage of diseased homozygotes. Malaria—especially if it is caused by the most virulent parasite, *Plasmodium falciparum*—is a very severe illness producing an anemia that lowers the body's resistance to other infectious diseases. For this reason it is especially lethal among children. Several different genes can increase malarial resistance; they are primarily found at high frequencies in those populations that have lived with some type of malaria for at least fifteen or twenty centuries. This is the amount of time it takes for natural selection to reach stable gene frequencies as high as those found in many populations that have been subject to serious endemic conditions of malaria.

We do not know how many genes experience heterozygous advantage, but it is one of the factors that make "racial purity" impossible: This type of natural selection will always maintain heterogeneity for a gene in a population until heterozygotes are at an advantage.

Genetic Variation between Populations

We have very few direct measures of evolutionary rates because we don't know the gene frequencies of past generations. However, we do know how genes vary in space, and there is a close connection between variation in time and space. On this basis it is clear that the rate of evolution differs greatly among genes.

If we knew the effective population size and the intensity of migration for the entire history of our species, and if we knew which genes were subject to natural selection, we could predict the distribution of genetic variation across the globe. The nature of drift would make this only a probabilistic prediction. In general, we would expect the same level of variation at any gene since the population size for every gene would be the same. Natural selection, when present, could reduce or increase the rate of evolutionary change. But different genes are subject to very different intensities of selection, and there are many genes that show no sign of it.

Other factors could limit the effect of drift. Because of migration, genetic exchange among populations almost always occurs, most often between closely neighboring villages. This migration (also called gene flow) tends to reduce genetic variation between villages. If it were extremely high, there would be no genetic differences between villages, nations, or even continents; but levels of migration that would wipe out genetic differences between populations clearly have not occurred. A high mutation rate could have an effect similar to migration, but the majority of genes that we have studied have a rather low mutation rate. We can usually recognize genes with a high mutation rate, because they have a higher number of alleles.

Among the genes that show the greatest geographic variation are the immunoglobulin genes (which produce antibodies, key elements in the body's defense against infectious disease). Because there is considerable geographic variation in the distribution of disease, it is no surprise that we find important geographic variation in the genes that defend us from these diseases. As a result, we might expect patterns very different from those produced by drift. But the great variety of infectious diseases, and the genes that protect us from them, also play the evolutionary game—bacteria, viruses, and parasites are constantly mutating to evade our defenses. The effect of this random arms race is very similar to drift in the sense that chance plays a major role in creating genetic variation of pathogenic organisms but is not affected by human population size to the same extent as are selectively neutral genes. It is therefore easy to understand why the results from an evolutionary analysis of immunoglobulins are similar to those obtained from the analysis of genes governed by effective population size (i.e., drift), except that their overall rate of change is greater.

The same goes for the HLA genes, an important and even more variable class of genes that are involved in our immunological individuality, and also in our immune defenses against infectious disease. The situation here is more complex. The HLA genes have many alleles; some of them are found everywhere, and others are restricted to particular regions. The most unusual pattern of variation (above all for HLA, but also for other genes) is found in the indigenous populations of South America, which generally show the greatest geographic variation. Almost all other populations in the world have a great variety of HLA alleles, but there is a single

HLA variant, rare or unknown elsewhere, that can achieve high frequencies in one or a few South American native populations. Neighboring populations frequently contain a completely different set of alleles. It is not easy to exclude the possibility that these elevated frequencies are due in some populations to natural selection, but it seems here that drift has also played a major role.

The problems of predicting evolutionary rates are, therefore, complex. A detailed study of a population can help us determine whether genetic variation is essentially random as a result of genetic drift or due to natural selection varying randomly (as for antibody genes and HLA), because we know how population size affects variation due to drift, and we can now easily increase the number of genes and individuals studied to reduce observational errors.

But some genes show very little variation from one population to another. In such a case, heterozygote advantage is probably stabilizing the gene frequencies, and therefore reducing their subsequent evolution. Homogeneity may sometimes be apparent. In malarial areas, for example, thalassemia is frequent. But molecular analysis has shown that in some geographic areas there are a great number of different thalassemia alleles, and this heterogeneity can be observed only by study at the DNA level. Often thalassemia alleles can provide information about ancient migrations.

The majority of genes have an intermediate geographic heterogeneity, between that of the highly variable genes involved in immunity and genes that do not vary at all. These are probably subject to a situation of heterozygous advantage common to all environments. The average level of variation of gene frequencies among populations is approximately in the range expected for selectively neutral genes and the particular population size we estimate for modern humans more than 10,000 years ago. Since then there have been significant innovations in food production. They generated a substantial population increase, causing a progressive freeze of drift. It seems therefore that genetic drift played a major role especially in early human evolution, and more recently only in special situations. Highly elevated or reduced variation for certain genes, however, must be caused by natural selection, which accelerates or retards their evolution.

Our task in reconstructing evolutionary trees would be greatly simplified if we could be sure that the rate of evolution—calculated as the mean of many genes (excluding genes subject to strong selection)—is approximately the same in different branches of the evolutionary tree. We have given an idea of the factors that can influence the rate of evolution. Do we have the ability to ensure that reality is as simple as our hypotheses?

We have seen in a table of genetic distances that Africa is genetically the most distant continent from all the others. In effect, the distance between Africa and the four other continents is 21.7 ± 1.7, almost double the distance between Oceania and the three other continents, 12.7 ± 1.4 each. This indicates that the difference between the means, 9.0, is well above the level of statistical error. The other distances are all much smaller. There is an excellent historical explanation for this result, which we shall see later.

In order to examine the problem of the constancy of evolutionary rates, we can look at the distances between Africa and the other continents: 24.7 with Oceania, 20.6 with Asia, 16.6 with Europe, and 22.6 with America. It is clear that the shortest distance is between Africa and Europe, followed by that between Africa and Asia. If the rate of evolution were truly constant, the four values would be identical (within the limits of statistical error due to small sample size).

The distance from Europe is anomalously low. North Africa is populated with Caucasoid people like Europeans, but we have made sure to eliminate these populations and are restricting ourselves to sub-Saharan Africa. The simplest explanation is that substantial exchange has taken place between nearby continents, probably via migrations in both directions.

Other comparisons should convince us that the proximity of the two continents contributes to their genetic similarity. Asia, for example, Africa's other neighbor, is genetically closer to Africa than is either America or Oceania. Comparing genetic distances between Oceania and the other continents, we have a similar situation: 10.0

with Asia, 13.5 with Europe, and 14.6 with America. The shortest distance is seen with the closest continent, Asia.

If the evolutionary rates of each continent were constant, the greater similarity of neighboring continents should not exist. But the deviation is not important. It does not need to result from different amounts of drift or selection in each continent. The underlying cause is rather the genetic exchange between neighboring populations, which changes genetic distances by reducing the distance between populations that have exchanged migrants. I conclude that even if evolution in the different branches of the tree is not completely independent, the deviations are not sufficient to destroy our conclusions. Migration is almost always limited to rather short distances. Nonetheless, it merits a digression.

Migrations Big and Small

Humans are constantly moving. Over most of our history, all of us have been hunters and gatherers, and most of us have been herders and farmers, though only in the last ten thousand years. Hunting territories were not very far from each other, and probably did not change hands too frequently. For the African Pygmies, these territories belong to a group (a hunting band) and each husband has the right to add his wife's territory to his own. For this reason, Pygmies seek to marry a woman from relatively far away, following a rule that, while extending his sphere of influence, has also decreased the likelihood of marrying a close relative.

Pygmies avoid marrying close cousins, but they don't keep track of more distant relationships. Hunters and gatherers require greater mobility than farmers, but according to current data the difference is not so great. As to herders, they often migrate over distances approaching 500 to 1,000 kilometers, although these seasonal movements are restricted over the years. It is almost never a random nomadism. These movements continue in parts of the world today, but they usually involve only a few shepherds and not whole groups. Certainly there have been other reasons to move around, usually without resettlement, to attend markets, festivals, and so on. Movement is also important for meeting potential spouses.

Marriage is a major reason for migration, since at least one of the spouses (more often the wife) must move to unite with the other. We have witnessed great changes in transportation this century, but movement was once limited, and was rarely greater than a day's walk before the widespread use of trains and other modern mechanisms of travel.

From a genetic point of view, the migration that matters most is one that causes a difference between the birth place of parent and child. This would include resettlement of one or both of the spouses at marriage, and any subsequent resettlement.

Data on the distances between birth places of a husband and a wife, which are the easiest migration data to collect, indicate:

1. an average distance of 30 to 40 kilometers for hunter-gatherers in tropical areas (these are much smaller than those for Arctic Eskimos, who have a very low population density);
2. 10 to 20 kilometers, on average, for African farmers with a low population density;
3. 5 to 10 kilometers for European farmers of the nineteenth century;
4. that during the second half of the nineteenth century, the average distance started increasing as a result of railroad construction.

These are modest levels of migration. By focusing on the distance between spouses, averages are small since most marriages occur between residents of the same village or town, sometimes separated by only a few blocks. This should not surprise us since people are more likely to meet their neighbors in the context of work, school, or recreation. Even in the smallest villages, most marriages of the rural Italian population occur between spouses from the same or the closest villages, and only rarely from distant villages.

These "minor" migrations of a family or an individual are behind the relationship between genetic and geographic distances (the "isolation by distance" we discussed in the first chapter).

Mass migrations are an entirely different phenomenon. Much rarer, they are very important in the history of our species. One type of major migration involves the deliberate settling of new territories. We call this *colonization*. There are several known historical examples: among them, the Greeks and Phoenicians of the Mediterranean and the European colonization of the Americas, Australia, and South Africa. There must have been many prehistoric colonization events as well. We will discover some of them in chapter 4.

In recorded history, colonizations have been well organized and were usually motivated by overpopulation. Previously, there must have been less organized migrations as well. Population growth can end in saturation, prompting migration. This may repeat in the new locale, the growth of which may lead to further cycles. We shall see in chapter 4 that these expansions leave a characteristic signature on the geographic map of genes.

The geographic study of genetic variation is very different from an approach based on evolutionary trees. It simultaneously solves and creates different problems. In the study of trees, one usually selects a small number of populations, usually located far apart, and tries to determine their historical relationships. Since all humans share a common origin, we can expect that a single or few populations have grown and begun to spread over the earth, reaching new continents and ultimately covering the entire earth. This type of migration, going from one geographic region to a remote, separate one, will create a discontinuity, the equivalent of a fission. If repeated, this process corresponds to the branching pattern of a tree. Thus, migration followed by separation of the mother and the daughter colony can cause differentiation; while migration between neighbors has an opposite effect, favoring homogenization.

But the process of colonization can be less abrupt, even at the beginning. And in later growth and development, neighboring populations are likely to have frequent, more or less reciprocal genetic exchange. This mixing can render the model of a branching tree inadequate in representing human evolution. In general, tree reconstructions are less useful than geographic or other more specific methods, but they help give a sense of the similarities of populations, and sometimes even help recognizing admixture.

An important control on the validity of an evolutionary tree is that all the genes or characteristics used point to the same result or at least have an explicable difference. If enough genes have been studied in widely separated populations and statistical tests are used to verify the stability of results, we often find strong support for a tree structure. Some individual branches may pose problems. Very short branches may be found for populations generated by admixtures, or affected by long lasting gene flow from neighbors. In these cases, the position of the relevant branch may also be shifted toward the center of the tree; for instance, capital cities which have experienced much immigration from all or most other parts of a country usually occupy a central place in a tree of the country. The length of a branch can also be affected by drift. A population that has had a very small number of founders, or later demographic bottlenecks, may show an inordinately long branch.

Even when independent samples of genes show substantial similarity, there are also differences that may be very informative. The genetic markers mostly used until now are proteins, the gene products rather than the genes themselves. The more recent markers employed in the direct study of the DNA present many advantages over protein markers, again, with only one drawback—they have been studied in only a few populations, while many protein markers have been examined in thousands of different populations.

Many difficult problems remain before we can satisfactorily resolve questions in human evolution through the analysis of living organisms alone. We shall devote the next chapter to a critical study of the problems that arise during comparisons of data from different genetic systems and archeological results that can help us reconstruct history by another means.

CHAPTER 3

Of Adam and Eve

Who Are Modern Humans?

Darwin was the first to point out that the great apes are our closest living cousins. Since the two most similar to us, chimpanzees and gorillas, live only in Africa, he concluded that we must have evolved there from ancestors common to us all. We now know that the last ancestor common to the chimp and humans lived about five million years ago. We must go back in time still further to arrive at the branch leading to the gorilla, a more distant cousin, and as far as thirteen million years to reach the point when we shared a common ancestor with the orangutan. Even so, and in spite of its long red coat, the orangutan still bears a surprising resemblance to our species. It lives in Southeast Asia, while our closer relatives all live in Africa. Fossil australopithecines, among our ancestors after the separation from chimpanzees, have so far been found only in Africa.

The first member of our genus, *Homo,* was *Homo habilis,* who appeared about 2.5 million years ago. *H. habilis* made crude stone tools and was completely bipedal. His brain was larger than that of his immediate ancestors and of the living apes, although it was still smaller than ours. There is complete agreement that *habilis* evolved in Africa, where *Homo erectus* succeeded him. The latter species was the first in our lineage to leave Africa and explore the rest of the Old World. Recent evidence suggests that *erectus*'s migration may have begun as many as two million years ago, not one million years ago as was previously believed.

With the arrival of *Homo sapiens,* about 500,000 years ago, we reach the cranial volume of modern humans. Many features of *sapiens* were still somewhat simian early on, and skulls like those of fully modern humans appear only within the last 100,000 years in southern and eastern Africa.

Some confusion was generated by the discovery of modern human skulls in the Middle East that were about the same age as the earliest African samples. Clearly, the Middle East is nearby and connected to Africa by land, but the find still raises doubts about the origin of modern humans: Africa or the Middle East? It turns out that Africa is the likelier source, given that other skulls found there proved to be earlier and transitional between older ancestors and modern humans.

The story gets more complicated. Neandertals, a branch of late *erectus* or of early *sapiens* that had been found only in Europe and western Asia in the last 200,000 or 300,000 years, also began appearing for the first time in the Middle East 60,000 years ago. The archeologist Richard Klein suggested that the first migration of modern humans from Africa to the Middle East did take place around 100,000 years ago but failed. A possible explanation is a local cooling of the climate. Since Neandertals had adapted to colder conditions in Europe, they may have migrated into the Middle East about that time and found it empty of modern humans.

Some paleoanthropologists long believed that modern Europeans are directly descended from Neandertals. But we have seen in chapter 2 that very recent analysis of fossil DNA (in particular, on the first specimen that gave the Neandertal name to the whole group) has shown that this can't be true. From these studies it became clear that Neandertals separated from the ancestral line some 500,000 years ago. Around 40,000 years ago they began disappearing rapidly, and are most probably totally extinct today.

Before the complete demise of the Neandertal, modern humans had begun expanding from Africa to the rest of the world. There are important reasons behind every major population expansion. In the case of modern humans, the most important have probably been technological innovations that improve food production, but discoveries facilitating transportation or climatic adaptations also contributed. One unique innovation, moreover, helped modern humans, born in Africa, to colonize the world.

The human brain grew continuously until the appearance of *Homo sapiens* about 500,000 years ago. Based on cranial measurements, growth of the human brain stopped at that point or shortly thereafter. In computer terminology, the "hardware" had improved, at least superficially, but that was not enough; the "software" also needed to become more powerful.

There is at least one major intellectual difference between us and our closest evolutionary relatives, the Primates. We can communicate with a much richer and more refined language than any other species. Chimpanzees and gorillas can learn to use only 300 to 400 words, and even that requires special effort and nonvocal communication, since they cannot articulate their tongues and pharynges to produce sounds comparable to ours. The vocabulary of an average human is at least ten or twenty times greater, and can be as large as 100,000 words or more. The great apes can use symbols to indicate simple things, but can understand these symbols only when somebody speaks the artificial languages devised by researchers who have conducted these remarkable experiments. However, they have a great deal of difficulty forming true sentences, and may be unable to develop grammar and syntax.

All contemporary modern humans use very complex languages. There are no "primitive" languages; the 5,000 or more spoken today are equally flexible and expressive, and their grammar and syntax are sometimes richer and more precise than that of the more widespread languages like English or Spanish, which have undergone some simplification over the centuries. All humans of normal intelligence can learn any language, provided they start at a young age. After the age of five or six, a child can almost never become perfectly fluent in a language, and the ability to learn it can completely disappear soon after that. After puberty, it is almost impossible to perfect the pronunciation of a second language. This is an excellent reason to begin foreign language instruction in elementary school, but few governments seem to have noticed this virtually absolute rule.

There is some indirect evidence that modern human language reached its current state of development between 50,000 and 150,000 years ago. As the archeologist Glynn Isaac has noticed, the Paleolithic cultures during this time show increased levels of local differentiation. This is reflected in the great number of names archeologists have given to the cultures of the period. Isaac postulated that this heightened variation in lithic culture, and the local differentiation of languages and dialects that most probably accompanied it, arose with an overall increase in language complexity. The possibility of communicating in a more refined manner, thanks to languages similar to modern ones, must have greatly aided our ancestors' ability to explore and colonize. Beginning perhaps 60,000 or 70,000 years ago, modern humans began to migrate from Africa, eventually reaching the farthest habitable corners of the globe such as Tierra del Fuego, Tasmania, the coast of the Arctic Ocean, and finally Greenland.

As I said earlier, other innovations over the last 100,000 years, such as improvements in toolmaking techniques, have been dominant factors in the most recent human expansion out of Africa. But advancements in navigation were possibly even more important. We do not have the remains of any boats or rafts more than 8,000 years old, since wood cannot survive so long, but we know that no fewer than seventy kilometers of water separated Australia from Southeast Asia, in four or five places. If modern humans could reach Australia, which was certainly occupied more than 40,000, and probably between 50,000 and 60,000, years ago, it seems likely that navigation techniques were available earlier. If so, modern humans did at least part of their colonization of Asia by sailing from Africa along the southern coasts of Asia, past Arabia, India, Burma, and Indonesia (not in one generation, of course, but over many hundreds of them). The seacoast and route is easier to traverse than an overland path, and it would not have required a change in diet from fish and shellfish, nor an adaptation to new climates.

Stages of Global Colonization by Modern Humans

The most crucial dates in modern human evolution are unfortunately beyond the range of the radiocarbon method, which has a limit of about 40,000 years. New alternative techniques can extend this range beyond 60,000 years. They have the advantage of being independent of carbon-containing material, and thus make it possible to assay the age of tools not made of bone or wood, but we have scarcely begun to use them and to appreciate their limitations. Taking account only of the earliest traceable human bone remains, there is excellent evidence that modern humans lived in southeastern Australia more than 30,000 years ago.

Archeology provides several dates for the first arrivals of modern humans on the various continents, and those dates can be compared to the genetic distances. The older the date of entry to a continent, the greater the time for the accumulation of genetic differences between the newly settled continent and the continent of origin. Thus genetic distances can be quite useful in determining when humans first reached a continent.

Modern humans seem to have reached Asia first. We saw above that they may have originally arrived in the Middle East as many as 100,000 years ago. If this first settlement was a failure because of climate cooling, there must have been a second, later settlement, but the first settlers may also have withdrawn farther south and southeast into Asia. How was the easternmost part of Asia reached? Was it set upon from the first Middle East settlements, or from East Africans traveling along the Arabian coast, and beyond India to

Southeast Asia? From here, two routes could have been followed: south to New Guinea and Australia, or north to China and Japan.

We know very little about the arrival of modern humans in East Asia—the only measured archeological date of human remains in China is 67,000 years ago, but this may be unreliable because of the method of measurement.

Europe was likely entered from western Asia and from North Africa, slightly before the disappearance of the Neandertals, about 43,000 years ago. The first entry to the Americas from northeastern Asia has been the most difficult to date accurately. Archeologically based dates range from 15,000 to 30,000 or even 50,000 years.

Many methods of genetic dating have been developed. Under the simplest hypothesis, the genetic distances between populations are proportional to the dates of first occupation of the geographic areas they occupy, and more precisely to the time elapsed since the separation between the pairs of populations being compared. The following table shows the dates of the first occupation of continents, on the basis of the archeological information given above, and the relevant genetic distances between pairs of continents. Genetic distances were calculated from blood groups and protein polymorphisms, and taken from the table already given in chapter 2. The last column shows the ratio between each genetic distance and the corresponding archeological date.

Migration	Genetic distance	First settlement date (thsnds yrs)	Ratio
Africa → Asia	20.6	100	0.206
Asia → Australia	10.0	55	0.182
Asia → Europe	9.7	43	0.226
Asia → America	8.9	15–50	0.59–0.178

The first three ratios between genetic distances and dates are reasonably similar one to the other, and the differences between the ratios are well within the error of measurement, confirming that genetic distance is approximately proportional to the time of separation of two populations. This is the same as saying that, in these data, the rate of evolutionary divergence among continents is at least approximately constant.

We do not have reliable dates for the initial settling of the Americas. Extreme values are shown in the last row of the table. It appears from the two corresponding ratios calculated in the last column that the latest date suggested by archeologists, 15,000 years, would be too recent, and perhaps the earliest one too long ago. Based on the average of the first three ratios, 0.205, we would estimate that the Americas were occupied 43,000 years ago (8.9/0.205 = 43). Notice that the distance between Asians and American Indians given here is probably too high, since it is based on all of Asia, although probably only the eastern portion of Asia participated in the colonization of the Americas. A more refined estimate would use the distance between East Asians and Amerindians, rather than that between the whole of Asia and Amerindians. From a table not given in this book, this distance is 6.6, generating an expected date of first settlement of America of 32,000 years ago (6.6/0.205 = 32).

Taking into account the difficulty of calculating a reliable arrival date to the Americas because of lack of agreement among archeologists, the dates of first occupation of the continents are in reasonable concordance with the genetic distances, but they require further refinement.

Nongenetic Data

From the outset, I was convinced that only strictly inherited traits like blood groups and proteins could provide satisfactory answers to questions of evolutionary history. For this reason, it was clear to me that external characters like height and other anthropometric measurements are not reliable, because they are influenced both by genes and by the environmental conditions of individual development; they change rapidly in response to factors such as nutrition and external temperature. Moreover, over time, the environment modifies the genetic basis of these characters via natural selection.

Characteristics subject to strong selection from the environment can tell us about the recent environmental conditions under which a given population may have lived. But we do not know how much time it takes to modify those characters. The best genes for evolutionary studies are, therefore, those that do not experience natural selection. Genes that have no function such as "pseudogenes" (duplicates of functional genes unable to produce a normal protein), or other DNA sequences that do not encode proteins and have no known function, are subject mainly to chance (genetic drift). Again, such genes are called "selectively neutral" and we prefer to use them in evolutionary studies whenever possible. Charles Darwin intuited this—he thought that the most useful characteristics for reconstructing history would be those that he called "trivial," being more easily subject to chance.

As I have said before, exceptions to this rule are the remarkably variable genes, like the HLA (a system of genes that control our genetic identity and help in immunity) or genes that make immunoglobulins (proteins that function as antibodies, protecting us from infectious diseases), that are among the most important in evolutionary studies. In principle they could also lead us astray when they are highly correlated with climatic or other environmental factors, which also influence the prevalence of certain infectious diseases. But chance always remains an important determining factor in their evolution, making them extremely useful.

When we were constructing our first evolutionary trees, however, it seemed important to conduct a parallel study using classical anthropometric characteristics. The idea was (and still is) that, in the search for solutions to a difficult problem, it may help to collect information from as many relevant sources as possible. If they provide different answers, the result must be explained. We collected anthropometric data of populations matching as closely as possible those for which we had genetic data. The resulting tree showed a few important differences from the genetic one. For example, Africans and Australians were very similar to each other and therefore grouped together in anthropometric trees, but in genetic studies these populations exhibited the greatest distance.

We were not happy, at first, about this, but it became clear that the cause of the anomaly is simply that anthropometric characteristics experience strong climatic selection. We know that skin color is largely determined by the sun's intensity. Sub-Saharan Africans, Australian Aborigines, and New Guineans have dark skin and have adapted similarly for other body traits, especially body measurements. They all live nearer to the equator than most other people. We also know that many other traits, such as overall size of nostrils, correlate in a physiologically understandable manner with climate, and therefore with latitude. Longitude does not show comparable ecological differences.

Anthropometric characteristics, including skin color, demonstrate the selective effects of the different climates to which modern humans have been exposed in the course of their migrations over the Earth's surface. They vary especially with latitude. By contrast, genes are considerably more useful as markers of human evolutionary history, especially migrations. They vary more with longitude.

The data that we used in our anthropometric study were derived from a large number of researchers, and as a result there was some heterogeneity in measurements between them. A beautiful, very detailed analysis was made by William Howells (1973) on a rich sample of crania. Based on multiple cranial measurements that Howells himself had made, he produced results very similar to the calculations we made from general anthropometric data. We were able to show that, after correcting for climate effects, and in particular eliminating the effects of general size (which is very sensitive to climate), we could improve the concordance between the craniometric and the genetic data.

In a second study of the same craniometric data, Howells (1989) tried to eliminate the effect of size by considering more specifically the shape of the skull. Shape is measured mostly using the relationship between the face and the crown. But shape is also very sensitive to climate selection. In very cold regions, modern humans show a strong reduction of the face in relation to the crown, which results in a major change in the form of the head. The use of shape did not change the conclusions, and the results of Howells's second analysis recapitulated the results of his first study. Characteristics strongly sensitive to natural selection by climate cannot provide a complete description of a species' evolutionary history but illustrate only a small part of it—that of environments occupied by different populations. The amount of evolutionary divergence accumulated is likely to be a measure of the time elapsed only when it reflects random changes.

Different Genetic Markers, Methods of Measuring Genetic Distances and Reconstructing Trees

After we began working on the reconstruction of human evolution using genetic trees, a multitude of new methods were proposed for calculating genetic distances. Similarly, many new methods for building trees were proposed. These various methods usually gave rise to trivial differences in results. In principle, it is easier to believe the historical validity of evolutionary trees reconstructed from genetic data if conclusions are not influenced in a significant way by the particular method used for calculating genetic distances or for reconstructing trees. Above all, conclusions should be independent from the genetic markers we use. If we do find that these variables have an effect, we must look for the reason, as we did when we found a discrepancy between genetic and anthropometric traits. Speaking of traits, however, we should remember that conclusions would inevitably depend on the number of traits we use: if too few are used, conclusions will fluctuate depending on the traits selected. This is a well-known limitation of all observations, which can be duly kept under control by appropriate statistical analysis.

As mentioned earlier, experience showed that the type of method used to calculate genetic distance does not have a major effect. But the method of tree reconstruction can. There are two major classes of methods: one is the standard statistical approach of making a specific evolutionary hypothesis and testing it against the data. But the most satisfactory (and also more complicated) method is called "maximum likelihood." The evolutionary hypothesis tested by this method is usually the simplest one: that evolution has a constant rate, the same in all branches, and that what happens in a branch is independent of what happens in the others. But one can change these hypotheses if it is reasonable to do so, as it sometimes is.

Another group of methods assumes that the evolutionary rate is the minimum possible. Some of the relevant methods are called "minimum evolution," and "maximum parsimony." One of them, called "neighbor joining," has some definite computer advantages over the others, which makes it very popular. There is no reason why evolution should be minimal, except that mutation rates are low and it has been proved that forcing evolutionary changes to be the minimum possible does not necessarily lead to the right conclusion.

The results we have shown so far are taken from the very large numbers of gene frequencies observed for blood groups like ABO, RH, and many other genes, mostly encoding enzymes and other proteins. A data bank we have collected includes almost 100,000 gene frequencies data on about 2,000 populations, published in the scientific literature since World War I. The trees we have used, and the geographic maps shown in the next chapter, are derived from them.

When we analyze DNA, we often abandon the study of gene frequencies in populations and instead examine individuals directly. Genetic distance between two individuals is simply obtained by counting the number of mutations making an individual different from another.

The markers that we introduce in figures 2A and B are from DNA studies. We know that DNA is the hereditary material. Remember it is made up of four types of nucleotides known by their initial as A, C, G, and T; and that the genetic information contained in DNA is entirely coded in its nucleotide sequence. There are more than three billion nucleotides in a single complete set of human chromosomes as found in a gamete (sperm or egg cell). Many readers know that DNA takes the shape of a double helix in which nucleotides come in pairs. The only possible pairs are AG, GA, CT, or TC, and therefore it is necessary to know the sequence of nucleotides in one helix only: in front of A in one helix, there can be only G in the other; in front of G, only A; in front of C, only T; and in front of T, only C.

If we take the DNA from one sperm (or egg) and compare it to the DNA of another random one, we find that there is on average one different nucleotide pair every thousand nucleotide pairs. There are therefore at least three million differences between the DNA in one sperm or egg and the DNA in another. All these differences originated by mutation, a spontaneous error made while copying DNA, which most frequently involves the replacement of one nucleotide by another of the four. New DNA is always a copy of the old, apart from mutations, which are rare. New mutations are therefore transmitted from parents to children. They accumulate in a population, and the mutation separating two different alleles found in a population may easily be tens or hundreds of thousands of years old.

It is possible to detect and count DNA differences between two individuals by sequencing all the nucleotides of a specific DNA segment, but the procedure is tedious and we now have many shortcuts to identify the presence of mutations.

The first method of studying changes in DNA became available in 1981 and uses the so-called "restriction." It requires a lot of DNA, and for this reason it became common to increase the amount of DNA that would be obtainable from a small amount of blood by transforming certain white blood cells—the B lymphocytes, which produce antibodies—so that they can reproduce continuously in laboratory culture. The procedure calls for infecting the cells with the Epstein-Barr virus (EBV), which induces them to divide without limit. This procedure has been nicknamed "immortalization." Naturally it is only a very specialized cell of an individual, not the whole individual, who has been immortalized. One can thus generate large amounts of DNA, which can then be used for a great number of tests. The procedure requires freshly collected cells, although cells frozen in liquid nitrogen can often be transformed much later. Even if, by polymerase chain reaction (PCR)—the enzymatic multiplication of

DNA in a test tube—one can produce large amounts of DNA from a single molecule, EBV transformation of B lymphocytes remains very useful, because the multiplication of DNA in vitro is never as precise as that produced by living cells.

With Ken and Judy Kidd of the Yale University Genetics Department, and anthropologist Barry Hewlett of Washington State University, in 1984 I started a program to produce these cell lines from a number of indigenous populations from throughout the world. The first collaborative effort was the generation of cell lines of African Pygmies of the Central African Republic and northeastern Zaïre. This was followed by other similar initiatives. In 1991 many of us proposed extending this program to a large number of human populations representing the entire human species. This larger plan became known as the Human Genome Diversity Project (HGDP), for which the U.S. National Science Foundation has begun making funds available. A pilot program, called the Biological History of European Populations, began in 1992 in Europe under European Community financing, and similar projects are underway in India, China, Pakistan, Israel, and elsewhere. At present, there is a growing collection of more than fifty populations, generated in seven laboratories. DNA produced from these cell lines will soon be distributed to research workers by CEPH, a French Center for the Study of Human Polymorphism, founded and presided over by Jean Dausset, Nobel Laureate and discoverer of HLA. CEPH has already given a fundamental contribution to human and medical genetics by promoting a global collaboration of world scientists that generated the human chromosomes genetic linkage maps, a major advance in the field. Some of these cell lines are also available to researchers through a cell culture facility maintained by the U.S. National Institutes of Health.

Figures 2A and B compare a tree obtained by the method of chapter 2 with one obtained by a minimum evolution method on nine populations by restriction analysis of DNA. Most of the populations appearing in the figures come from our collection of transformed cell lines. Two African Pygmy populations are included: one from the southwest of the Central African Republic near the village of Bagandou, which I visited again for this purpose in 1984, and the other from a trip I made in 1985 to the Ituri Forest in Zaïre. The Mbuti Pygmies from Ituri are the shortest while those from the Central African Republic are taller, probably because they mixed as much as 75 percent with nearby Bantu and Sudanic villagers, but it is also possible that another mutation is involved in decreasing their size. The Mandenka are Senegalese samples collected by André Langaney of Geneva and his colleagues. The European samples were obtained by Howard Cann from a Mennonite population in

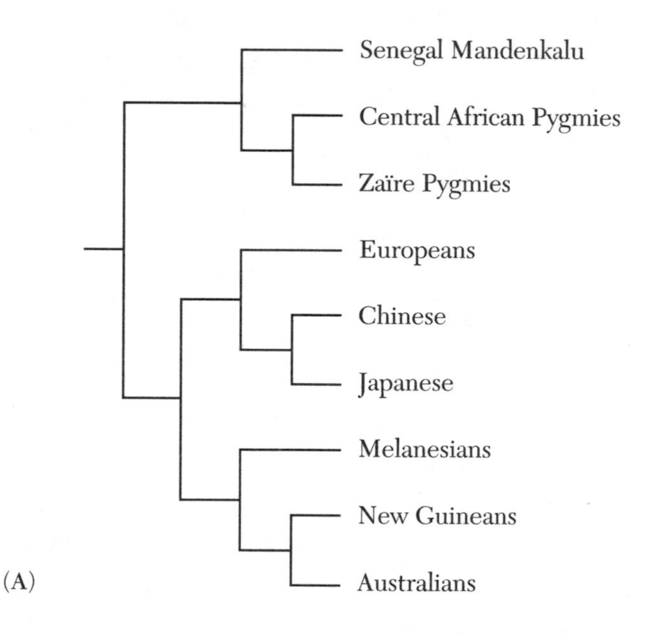

Figures 2A and B. Trees generated for nine populations using 78 DNA restriction markers, with two different methods: (2A) tree assumes a constant rate of evolution (average linkage, maximum likelihood); and (2B), by a technique called "neighbor joining," assumes the minimum evolution necessary to generate the observed distances. Both hypotheses have serious limitations. The tree in B is drawn with distances proportional to those calculated, and tries, with partial success, to fit a world map. The numbers represent the value calculated for each tree segment. It is clear that the European segment is far too short. (Data and figure from Poloni et al. 1995)

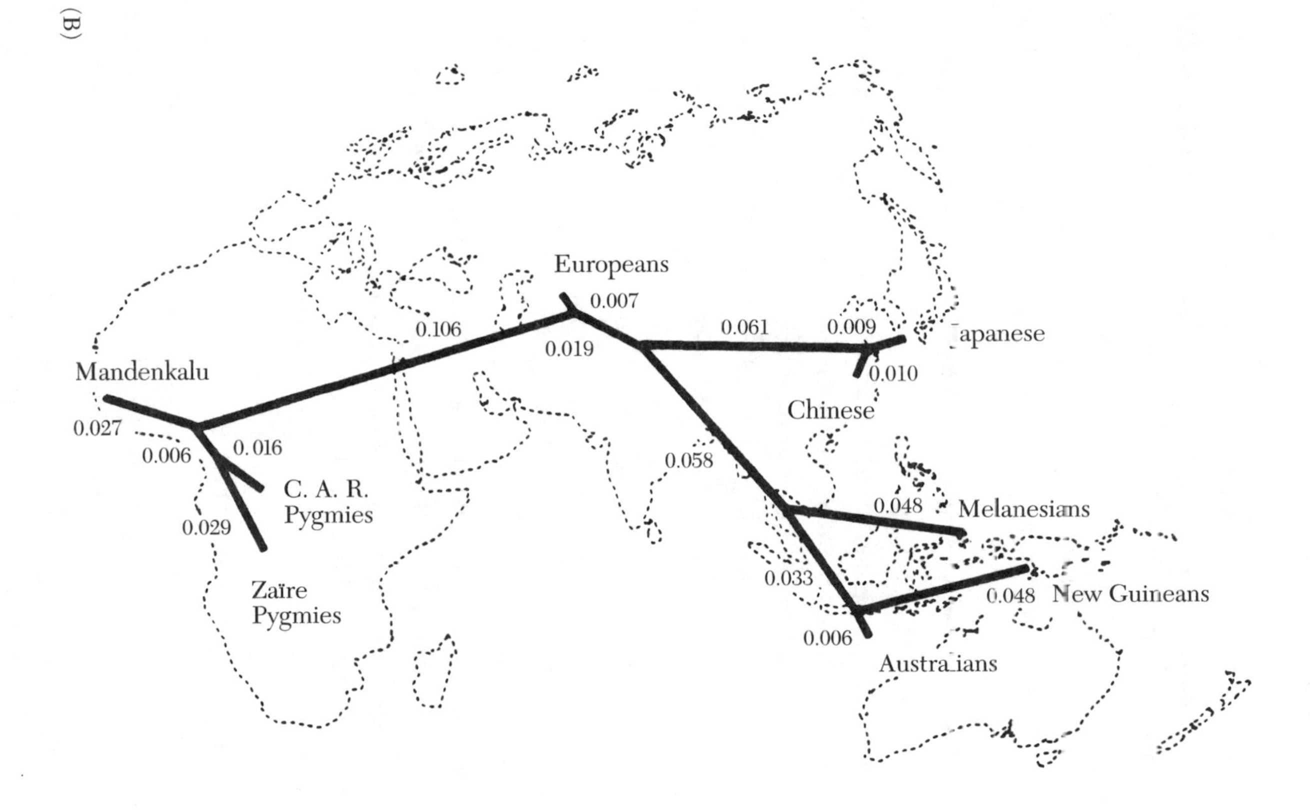
Europeans
0.007
0.106
0.019
0.061
0.009
0.010
Mandenkalu
Chinese
0.027
0.006
0.016
C. A. R.
Pygmies
0.029
Zaïre
Pygmies
0.058
0.048
Melanesians
0.033
0.048
New Guineans
0.006

(B)

California, originally from Germany and Great Britain. Chinese subjects (mostly southern Chinese) and Japanese were all born in diverse parts of Asia, but now also live in California. A single Oceanian population is represented by Melanesians from Bougainville Island, whose blood samples were collected by Jonathan Friedlaender of Philadelphia. The Australian Aborigines and New Guineans come from various parts of those regions. In independent studies, we tested indigenous Central and South American populations, which are placed in the tree exactly where we would expect, based on work with other markers.

There are today many types of DNA polymorphisms other than those detected by restriction enzymes. They all contain potentially more information than the classical markers we relied on in the previous chapter, but there has not been time yet to accumulate data on a sufficiently large number of populations, as there exist from the earlier work with blood groups and proteins. The different DNA markers tested so far confirm the earlier results, and in some cases they have already allowed scientists to push the analysis further.

The agreement between different types of markers and the two methods of testing is high, but not complete. All world trees place the earliest split between Africans and non-Africans, which is expected given that all modern humans originated in Africa. But if you assume a constant evolutionary rate, the next split tends to separate Oceanians from the rest of non-Africans, while minimum evolution locates it between Europeans and the rest. There is also another, startling difference between the two trees. With minimum evolution and related methods it is easier to observe length differences among the various branches than is possible with methods assuming constant evolutionary rates. This is not surprising because, unlike the latter methods, minimum evolution places no constraint on the relative length of tree branches. A striking difference between trees obtained with the two methods is that minimum evolution places Europe—and to a lesser extent, East Asia—on a very short branch departing near the center of the tree. The agreement between markers, despite the disagreement between evolutionary trees, compels us to seek an explanation for the short European branch observed in minimum evolution trees, and for its unexpected central placement. This is at odds with the archeological information on settlement times, according to which the European branch should depart from the tree as represented in the constant evolutionary rate tree.

Length of Branches in Trees

When we see different branch lengths in a tree reconstructed by the minimum evolution method, the simplest hypothesis is that a short branch is the result of a slow rate of evolution in that region and a long branch that of a locally rapid evolution.

Two major evolutionary factors—drift and natural selection—can change in important ways from one place to another. Drift affects all genes. For a particular population, drift has about the same intensity for each gene, as a function of the population's "effective size." As opposed to census size, effective size refers only to those people who reproduce, the generation intermediate between those too young or too old to do so—approximately one-third of the total population. Natural selection, on the other hand, is free to change any gene, in any population, at any time. It is likely that only a few genes are under strong natural selection for any long period of time. The genes for which we see maximal differentiation and that appear to be under the effect of selection varying randomly in space and time (like the HLA or antibody genes) produce evolutionary trees similar to those obtained with genes subject to drift. It is, therefore, unlikely that a short or a long branch can be traced to differences in natural selection.

So, is drift responsible for variation in branch lengths? Demographic information can help to evaluate this possibility. In the case of a small island that has not had any recent immigration, drift might explain a long branch. Numerous examples exist. Easter Island is very far from both the South American coast and the other Polynesian islands. The demographic history is known in broad outline, and it shows that a severe population bottleneck occurred in the eighteenth century. As a result, Easter Islanders have a longer branch than other Polynesians. Sardinia is another example. It is the most isolated of Mediterranean islands and its history reflects long cultural isolation. To a lesser extent, this is also true of Iceland. Although this island is relatively remote from other lands, it is less genetically different from the rest of Europe than Sardinia, but we know that it was settled relatively recently, in the ninth century, by a rather large number of colonists (about 20,000). The information on which these and many other similar statements are made can be found in *The History and Geography of Human Genes,* listed in the bibliography.

Small gene flow caused by geographic isolation and small population size are not the only reasons for long branches. Cultural factors have limited the exogamy of groups such as Basques, Jews, and Eskimos, who tend to marry mostly within their own group. In the cases of high "endogamy," due to geographic or cultural isolation, long branches are possible, especially if the group is small. A small population size and a reduction or total lack of marriages outside the group can lengthen branches in an evolutionary tree.

Short branches have the opposite cause: large population sizes, which reduce drift, or elevated levels of genetic admixture. When marriage to another group is frequent, the original ethnic identity is gradually lost. Admixture is frequent when migration puts two groups in proximity. The migration of Africans brought to America as slaves has resulted in mixing between Blacks and Whites, and Blacks and Native Americans. In some areas mixing between all three groups has occurred to produce "triracial isolates." Black people who have partial White ancestry are generally still classified as Black in North America. Black Americans have received a considerable input of genes from Whites. Studies with genetic markers indicate an average of 30 percent of White admixture in the Black population, the frequencies varying between approximately 50 percent on average in the northern United States and 10 percent in the South. Over the three centuries during which Blacks and Whites have lived together in America, 5 percent of White genes would have to enter the Black population per generation to reach an overall level of 30 percent.

At least three major instances of gene flow occurred on the continent of Africa itself (although there are probably many examples that have not been studied). In North and East Africa there must have been many opportunities of admixture between Blacks and Whites. In the north there is a preponderance of White genes while Black genes are more predominant in the east (60 percent, on average). Along the Nile, Blacks have lived in the south and Whites in the north for at least the last 5,000 years. Contact between Ethiopians and Arabs occurred very early; at a later time, from about 1,000 B.C. until fairly recently, a mixed Arab-Ethiopian empire ruled first from capitals in Arabia, then from Aksum in Ethiopia.

A legitimate question, which is difficult to answer today, is where white skin arose. It is not impossible that it arose in Africa itself, maybe in the north, or both in the north and east. We do not know enough about the genetic determinism of skin color, except that there must be at least three or four different genes.

Let us discuss, for a moment, the very short branch of Europeans in the minimum evolution tree of figure 2B. From what we said before, it could have one of two explanations, or both: no drift at all (i.e., a very large population throughout), or an admixture. The first explanation is extremely unlikely. The shortness of the European branch and its central location would mean that Europeans did not evolve but remained almost unchanged from the humans who lived 100,000 years ago. But our recent knowledge of the effects of the last glaciation suggests that the population of northern Europe was reduced considerably between 25,000 and 13,000 years ago, and after the end of the glaciation around 13,000 years ago the continent was resettled starting from its southern shores. This would have lengthened, not shortened, the European branch.

The second explanation is that Europeans are the result of genetic admixture, most probably resulting from migrations from the two neighboring continents, Africa and Asia. A calculation of its genetic consequences fits exactly the data, as shown by Bowcock et al. (1991). If we sought to determine the exact composition of this mix, it appears that Europeans are about two-thirds Asian and one-third African. When would this mixture have occurred? The data suggest a rather early date, on the order of 30,000 years. How can we further test such an explanation? It is a challenge that the forthcoming data on new DNA markers may well answer.

Arthur Gobineau, the nineteenth-century French diplomat and intellectual who wrote the very popular "Essay on the Inequality of Human Races," which contributed to the rise of German racism, would die of rage and shame at this suggestion since he believed that Europeans (especially those from central Europe, who are at the genetic center of Europe) were the most genetically pure race, the most intellectually gifted, and the least weakened by racial mixing. His persuasion that race mixture is the cause of degeneration became very popular, but is at odds with everything we know today.

We must consider a third, artifactual reason for the central location and shortness of the European branch in minimum evolution trees. Almost all of the classical and DNA polymorphisms that we have studied so far were first identified in blood samples from Europeans or their descendants in North America. Most markers were designed for use in the study of linkage—a technique that allows the chromosomal location of disease genes to be determined. The polymorphisms that best suit this purpose have an equal frequency for all the alleles, and were therefore selected preferentially. Europeans were the source of almost all the genetic polymorphisms that we have studied so far, irrespective of whether DNA or proteins were used. Could this have artificially placed Europeans in the center of our trees? The answer seems to be yes, although a deeper analysis reveals that this explanation can be responsible for only one part of the phenomenon.

It is true that the study of migrations, of which we shall speak more in the next chapter, has shown that an important portion of European genes derives from the Middle East. It is also true that the Huns, an East Asian population, arrived in France and Italy around A.D. 450. And it is true that the Turks made it to the Austrian border at the end of the eighteenth century. But the geographic distribution of genes in Eurasia assures us that these incursions had few genetic consequences. It is more likely that the intermediate position of Europe between Asia and Africa is the result of more ancient admixture than the last two.

Mitochondrial DNA and the Story of "African Eve"

The study of mitochondrial DNA (mtDNA) has generated much enthusiasm, in part because of how easy it is to work with. Mitochondria are small organelles found in the cells of all eukaryotic cells (cells of higher organisms, which, unlike bacteria, have a regular nucleus). In a single cell there are often several thousand, even tens of thousands of these organelles. Their job is to generate the cell's energy supply by using oxygen to liberate the energy contained in organic molecules (primarily sugars). The transmission of mtDNA appears purely maternal. It is possible that one or a few mitochondria from the sperm enter the egg at conception. This has been observed in mice and could in exceptional cases occur in humans but only very rarely, and in any case, the paternal mitochondria would be greatly outnumbered by the mother's. It appears that mitochondria are the vestiges of bacteria that entered a eukaryotic cell and became symbiotic with it more than one billion years ago. Today the symbiosis is obligatory for both the host cell and the mitochondrion. The mitochondrial genome is very short—a bit longer than 16,500 base pairs, which is vastly shorter than the three billion nucleotides of the nuclear genes. It contains genes that code for a few proteins and certain specialized RNA molecules. The most important genes generally vary little from one individual to another or even from one species to another. In most cases their variation would be incompatible with life itself. Mutations in mitochondrial DNA are, on average, at least twenty times more frequent than in the nuclear genes. The mutation rate is even higher in a short segment called the D-loop, which has been the object of most evolutionary studies. This elevated variability, although restricted to a small part of the molecule, helps certain evolutionary studies. In particular, claiming the likely extinction of Neandertal was made possible by analysis of the D-loop. In fossil bones, DNA is usually highly fragmented and very difficult to study, but the presence of many copies of mtDNA per cell and the fact that the original Neandertal remains were subject to a somewhat lower temperature were of considerable help in reaching this important conclusion.

Several laboratories, including ours, have shown that mtDNA gives similar, and sometimes identical, results to the autosomal markers we have used. The most complete study of mtDNA was done by the late Allan Wilson and his colleagues at UC Berkeley. They were the first to sequence the D-loop of a number of individuals from all over the world. Several years ago, I was surprised by a call from *Vogue* magazine requesting an interview about the birth date of "African Eve," which scientists had just dated to 190,000 years ago. Journalists knew earlier than I about the work being done in Wilson's Berkeley laboratory, fifty miles from mine.

Wilson was working on an application of the "molecular clock." If one can count the number of mutations that differentiate two living individuals, and identify when their last common ancestor lived, one can construct a "calibration curve." Either proteins or DNA should provide the same result. A well-known protein molecule, hemoglobin, was the first used for this approach by Emil Zuckerkandl and Linus Pauling, in the sixties. Our information about possible dates for the last common ancestors of pairs of individuals living today is better now than it was then. The most useful dates are linked to catastrophes like the fall of a meteorite near the Mexican coast around 63 million years ago. This event opened the crater of a volcano, causing a major eruption that obscured the sun and altered the climate so drastically that several groups of animals, like the dinosaurs, died, and others, like several orders of mammals, began to prosper. Counting the number of mutations separating, say, cattle and humans, whose last common ancestor lived probably a bit further back than that disaster, supplies one point from which to build the calibration curve linking that geological date and the number of mutations separating a cow and a human. Ideally one would like to have many different dates and corresponding mutation counts, each pair generating a point by which to build the calibration curve. (Actually, one point is sufficient because the theoretical shape of the curve is known from mathematical theory, but such a procedure is obviously less reliable.) Knowing the number of mutations separating chimps and humans, and comparing that number with the number separating cows (or other mammals) and humans, it became possible to establish that the separation of chimps and humans is about five million years old. This date could then be used, counting the number of mutations separating Africans from non-Africans and comparing it with the number separating chimps from humans, to establish the birth date of the so-called "African Eve." According to that first estimate, the woman from whom all modern human mitochondria descend lived about 190,000 years ago (with a probability interval of 150,000 to 300,000 years). As we shall see, this first attempt was not so bad.

While calling this woman Eve attracted a good deal of publicity, it was wrong and gave rise to much misinterpretation. Many scientists believed—and perhaps some continue to believe—that genetic data suggest there was only a single woman at that time, whom it was natural to name Eve. Because these mitochondrial data, like all other genetic data, indicated an African origin for modern humans, it was possible to call her African Eve. But it is clear that many women lived throughout that period. Their mitochondria, however, did not survive. "African Eve" is simply the woman whose mitochondria were the last common ancestors of all surviving mitochondria today.

Another frequent mistake is believing that the birth date of this woman was simultaneous with the first migration of modern humans out of Africa. In fact, it must have preceded it. The origin of a mutant gene that is the last common ancestor of a gene or of a DNA segment and the separation of populations are different events. The second event, the actual division of populations (e.g., the leaving of Africa by parties of modern humans settling in Asia) is later, possibly even much later. The same confusion has arisen for various other genes unrelated to mitochondria.

African Eve has caused great controversy in the scientific world. Many scientists have criticized both the date and the interpretation of its significance. I will not go into detail about criticisms of Allan Wilson's work and conclusions, since recent work from Japan confirms his results and provides a better estimate of the birth date of mitochondrial "Eve." Wilson's studies were limited to a small fraction of the mitochondrial DNA. Satoshi Horai and his colleagues have studied the complete mtDNA sequence of three humans (an African, a European, and a Japanese) and have compared these to four primate sequences (Chimpanzee, Gorilla, Orangutan, and Gibbon). They established "Eve's" age to be 143,000 years with a rather narrow confidence interval. The separation between Japanese and Europeans occurred much later, although again the branches refer to mutations in mtDNA, not separations between populations.

Adam

Should there be an Adam, to complement Eve? Yes, but we cannot expect him to have been born at a similar time and place. The processes of paternal and maternal transmission took place independently, and the only thing we can expect to be common to Adam and Eve was that they both lived in Africa, though not necessarily in the same region.

The key to finding Adam was the Y chromosome. Humans have 23 chromosome pairs, and as in most other living organisms they receive one member of each pair from the father, and one from the mother. One looks at chromosomes in a cell that is dividing, because then all chromosomes, normally very long and thin threads, are compacted into short rods. Each chromosome has a specific size and shape, and members of the same pair are identical to each other, with one exception: the sex chromosomes. These are two, called X and Y, the X being of average size relative to the other 22 pairs, and the Y being one of the shortest. Females have two X chromosomes, and males an X and a Y. It is therefore possible to distinguish the sex of an individual by simply looking at its chromosomes.

It is the Y chromosome that makes a male a male. A son receives an X chromosome from his mother and a Y from his father. Y chromosomes pass from male to male without end, and a mutation in one male will be found in all its male descendants.

The first single-nucleotide mutation of the Y chromosome was found in an African male. It took a laborious search. Until then, several laboratories had failed finding any such variation. No shortcuts proved useful, only brute force: we sequenced, in their entirety, seven DNA segments on many individuals from all over the world, before we found the first variant. The colleagues in my laboratory were not happy about having to endure this tiring procedure. I was away for some months, and when I returned I was in for a surprise: two of them, Peter Underhill and Peter Oefner, had developed a new method that helped them locate mutants more easily than ever before. In less than three years they accumulated some 150 new polymorphisms, with which they made a beautiful tree of Y chromosome variation, starting with orangutans, gorillas, and chimpanzees, and showing with greater clarity than ever before that the continents were settled by Africans in the expected order. Modern humans appear first in Africa, then in Asia, and from this big continent they settled its three appendices: Oceania, Europe, and America. By now, this story keeps repeating itself with any genetic system. As to the birth date of Adam, it is very similar to that of Eve: 144,000 years ago. But the similarity is superficial: both dates are affected by a statistical error of more than plus or minus 10,000 years. Even more important than proving that the origin of modern humans out of Africa holds true for males as well, the Y chromosome research has helped develop a new method of detecting mutants that can be applied to any chromosome. It is also proving very useful in one special branch of genetic variation research: that of genetic disease, that is, medical genetics.

There was one other coda to this Y chromosome study, much of the merit for which goes to Mark Seielstad, who helped translate this book into English while still a Ph.D. student. Y chromosome mutants have been found to be very highly clustered geographically, more than those of other chromosomes, or even of mitochondria. In other words, men move very little genetically. The old statement from Verdi's Rigoletto, "La donna è mobile," turns out to be true, though not at all in that old, frivolous sense: rather in an entirely new, genetic one. Most people find this hard to believe, as we are used to the idea men are the ones always on the move. That may still be true, but it's another story. Even when the anthropologist Barry Hewlett and I measured the geographic mobility of male and female African Pygmies, we found that the "exploration range" of males was on average about twice that of females. But for *genetic* mobility, what counts most is where people settle for marriage, and on average it is women, more often than men, who change residence to join their spouses. In times past, and still among some South American tribes now, when women were scarce it was usual to kidnap them from near tribes or villages, which made women even more mobile genetically in a more brutal way. The difference in genetic flow of males and females can help clarify ancient migrations. It introduces the possibility of duplicating observations, though with a different meter.

The Importance of Stuttering

Modern molecular genetics has already produced many discoveries. One of the most surprising is that the human genome (and those of almost all other species) contains large number of "repetitive" DNA, that is, sequences of nucleotides that are repeated, usually in tandem. Some, called "microsatellites," comprise very short repetitive sequences of two to five nucleotides. The most prevalent motif contains only two nucleotides, cytosine and adenine (C and A), and the DNA segment is therefore CACACACACACA . . . a sort of stuttering. Errors often occur here when DNA is copied so that the number of repetitions either increases or decreases in the new gene. Usually only one repeated unit is gained or lost at a time. When the mutation rate is high, we find many different numbers of repeats (e.g., 21, 22, 23, 24, and 25 repeats). A heterozygote will have two different forms, for example 22 and 25, coming to him from father and mother. These repetitive sequences (microsatellites) are numerous in the genome, and each one of them can serve as a genetic marker.

Many laboratories have been engaged in finding and mapping these repeats. The work of the French laboratory Généthon has been among the most productive; 5,264 isolated microsatellites were made available to all laboratories and have played an important role in generating the current map of human chromosomes. Microsatellites seem to be randomly scattered throughout the genome, and on average there is one roughly every 50,000 nucleotides. They have been most useful in locating hereditary disease genes, functioning as presumably harmless markers. A few of them, however, have quite unexpectedly turned out to be the culprits in important genetic diseases.

An interesting evolutionary application of microsatellites is a method of "absolute genetic dating." This method allows us to date population separations, impossible through other genetic methods.

We have seen that standard genetic methods estimate the last common ancestor's birth date, approximating only upper limits of population separations. Archeological dates hint at the first arrival of new settlers. These are often underestimates, because it is very unlikely to find evidence of the original settlers in the archeological record. The real date lies between the genetic and the archeological one, but the latter is likely to be closer.

Then there is the molecular clock method: it requires reliance on at least one other past event whose date is precisely known. Very few such events exist, and the nearest and most useful for our purposes, the separation of chimpanzees and humans, can be only approximately dated, with a 20 percent margin of error.

Microsatellites may provide an alternative. If we can ascertain the mutation rate, we can count the number of mutations separating two species and calculate their time of separation. Unfortunately, our estimates of mutation rates are questionable. Microsatellites grant an exception, because their mutation rate is so high (somewhat less than one per thousand) that it can be evaluated without excessive difficulty. The pattern of microsatellite mutation is a little complicated, because mutations happen in both directions (repeats can increase and decrease), and the change is not necessarily limited to one repeat at a time. Fortunately, Généthon has made an excellent analysis of the mutation rate and pattern on its 5,264 microsatellites. In an earlier analysis, in which the pattern of mutation was considered to be simply that of addition and subtraction of repeats one at a time, we obtained a value very close to that of mitochondrial Eve. But consideration of the observed mutation pattern, in particular the frequency with which more than one repeat is added or lost, decreased the estimate of the first migration out of Africa considerably, and brought it down to 80,000 years, very close to the archeological estimate. We are currently accumulating further data on more microsatellites and hope to publish a fairly accurate evaluation of this important date, which is central in the evolution of modern humans.

We have called this method absolute genetic dating, because it does not rely on paleontological dates, which are scarce and rarely reliable. Neither, then, does it consider the very approximate calibration curve on which the so-called molecular clock is based; even its theoretical shape, based on a shaky hypothesis, might be challenged.

All genetic dating methods that rely on mutation rates are independent of paleontological dates, and in this sense they are "absolute." They are of course as good as the available mutation rate estimates. Those provided for microsatellites by the Généthon group are very good, but they have been calculated for only a very special group of microsatellites (CACA . . .). This value is now frequently used for other microsatellites, but really there is no good evidence that this extrapolation is permissible. Application of mutation rates of other genes, in particular single nucleotide polymorphisms ("snips"), is not satisfactory. They are extremely low, on the order of 1 in 100 million per nucleotide per generation, or less, and have never been directly estimated by direct counting. The existing estimates are average values for poorly known genes. The only sure thing is considerable variation from nucleotide to nucleotide, and probably also from DNA region to DNA region. There should be some improvement of this situation once the sequencing of the whole genome (the Human Genome Project) is finished, and energy and machines currently tied up in the project become available.

Accurate knowledge of mutation rates is necessary for a serious application of absolute dating methods to evolutionary rates. The standard example of the power and weaknesses of the absolute dating methods is the use of radiocarbon (^{14}C) by archeologists for dating carbon-containing materials. The calculations use the rate of disintegration of radiocarbon, which is very well ascertained and stable. The method is absolute in the sense that it does not, in theory, require calibration from other sources of information. But there is at least one other important hypothesis that must be correct for radiocarbon dating to be acceptable: that the amount of radiocarbon available to plants in the atmosphere has been constant through time. This basic hypothesis was checked by comparing radiocarbon dates with other measurements of time. Tree rings that provided dates for the last 10,000 years against which to compare clearly showed that corrections to standard radiocarbon dates were necessary.

The genetic method of dating is based on the hypothesis that mutation rates are constant, and this may require further testing. One kind of test could be to measure mutation rates in people living in very different environments.

Science proceeds by successive approximations. The first time that the speed of light was measured, in 1675, there was an error of about 30 percent (200,000 km/sec). In 1732, a second measure gave 313,000 km/sec. Today we know it with an error of less than a meter. Almost all genetic results agree that modern humans arose in Africa and spread to the rest of the world in the last 100,000 years. Exact dating and routes will require further work, but new tools are becoming available.

I started thinking about reconstructing phylogenetic trees as a means to understanding human evolution in 1951, and I have since grown aware of the oversimplification that they make. Mathematical representation is inevitably simplistic, and occasionally one has to be brutal in forcing it to suit a reality that can only be very complex. And yet, there is a beauty about trees because of the simplicity with which they allow you to describe a series of events, like the differentiation of human populations. But one must ask whether one is justified in simplifying reality to the extent necessary to represent it as a tree. When Anthony Edwards and I started trying to fit trees to real data, I was aware of an alternative method, principal components analysis, that allows for a more faithful description of the data, and is always worth trying jointly with trees. It does not reconstruct a simple history like a tree does, and in fact it does not give a history at all, but it represents the whole set of data in a very simple graphical way, and reveals latent patterns, if they exist, in the mass of apparent nonsense that the original data seem to be at the beginning. It was therefore convenient to use both methods side by side.

Principal components analysis had been invented in the thirties, but had been applied only a few times, because of the staggering amount of arithmetical work it requires. Before the invention of computers, very few scientists were sufficiently determined to carry out such an enormous number of computations. To use a concise description very unfair to the non-mathematical reader, it simplifies the "data matrix," formed by the frequencies of the various alleles of many genes, observed in many populations, by calculating the eigenvectors of a few of its leading eigenvalues. It is difficult to explain it to non-mathematicians, other than by saying that it reduces the number of dimensions with which one can represent the data, with a minimum loss of information.

A classical application of principal components analysis shows how one can use distances between all possible pairs of cities of, say, Europe by car, train, or plane, in kilometers or in duration of transit, or all of them together, to draw a map in two dimensions that automatically reconstructs the geography of European cities with very good approximation. Remembering that there is a strong correlation between genetic distances and geographic distances it is therefore not at all surprising that by applying principal components to the genetic distances between all possible pairs of world populations, one can reconstruct a map of the world. There is inevitably some distortion, because genetic distances cannot be perfectly proportional to geographic distances by sea and land. Crossing oceans has been more difficult than crossing even vast tracts of land, at least until transoceanic navigation became easy, and the data we use reflect movements of indigenous peoples before that time.

When Anthony Edwards and I were calculating the first tree, we also produced the first principal component map of the same data. At the time, 1962, there were no packaged computer programs for doing the work. Anthony actually reinvented principal components analysis and I felt sorry to have to tell him I knew that it had already been invented. The two methods, trees and components analysis, are complementary. The first is more informative on history, and the second on geography. Using both at the same time on the same data can provide a synthesis of the two approaches.

In the next chapter we will use principal components again for a different, very specific geographic application. It may be useful to show here how one can represent data from 42 populations from all over the world for over 100 genes, using just two or three dimensions. Paolo Menozzi, Alberto Piazza, and I collected data from a survey of at most 100,000 gene frequencies of protein polymorphisms in about 2,000 world populations, the basis for an analysis that appeared in *The History and Geography of Human Genes.* The 2,000 populations were clustered in 42 groups, using criteria of geographic, ethnic, and linguistic similarity for pooling them. These 42 populations will be used for generating a tree comparable to that of languages on page 144. Here we show in figure 3 the analysis by a method very similar to principal components, called multidimensional scaling, which improves somewhat the efficiency of information recovery. We reduce the more than 100 genes to just two dimensions or axes, which still recover more than 50 percent of the total information supplied by the 100 genes. The vertical axis is the first and more important of the two. Like the tree, it separates Africans from the rest of the world. This is in accordance with the first split of practically all trees, which draws the same separation. The graph shows that the six African populations are scattered more widely than are all other thirty-six populations from the rest of the world, indicating they have been isolated more than all other populations. But they still form a cluster, from Mbuti Pygmies at the top to the East Africans who are closest to the rest of the world populations. This may mean that some of the Africans who went first to Asia were East Africans, but it may also mean that there has been considerable gene flow in later times between East Africans and Arabs. The geographic vicinity of these two regions agrees with both explanations, and it is difficult to weight them on the basis of these data alone. One-seventh of the African population, Berbers from North Africa, falls into the Eurasian cluster. Again there is the problem of distinguishing between the hypothesis that Berbers originate from an admixture of North Africans with Europeans (and also with people from the Middle East), and a second one that Berbers are direct descendants of the North Africans, a fraction of whom settled Europe. The two hypotheses are not mutually exclusive and might both be true. Hopefully DNA markers will give some clue for choosing between these hypotheses, and other possible ones, or indicating their relative roles if they all share some part of the truth. Incidentally, the fact that Berbers fall in the Eurasian cluster should not be taken as sufficient evidence for classifying them as having originated in Eurasia. More data might give a better picture, and lead us to a different conclusion. It seems almost certain that every population is made of genotypes that originated in more than one continent. We will eventually be able to reconstruct the origins of our genes in terms of a long and complex history.

The more disparate scatter of African populations is expected, as most of the history of modern humans took place in Africa. There was, thus, more time for differentiation into more diverse groups. We can visualize our evolution as having taken place in very

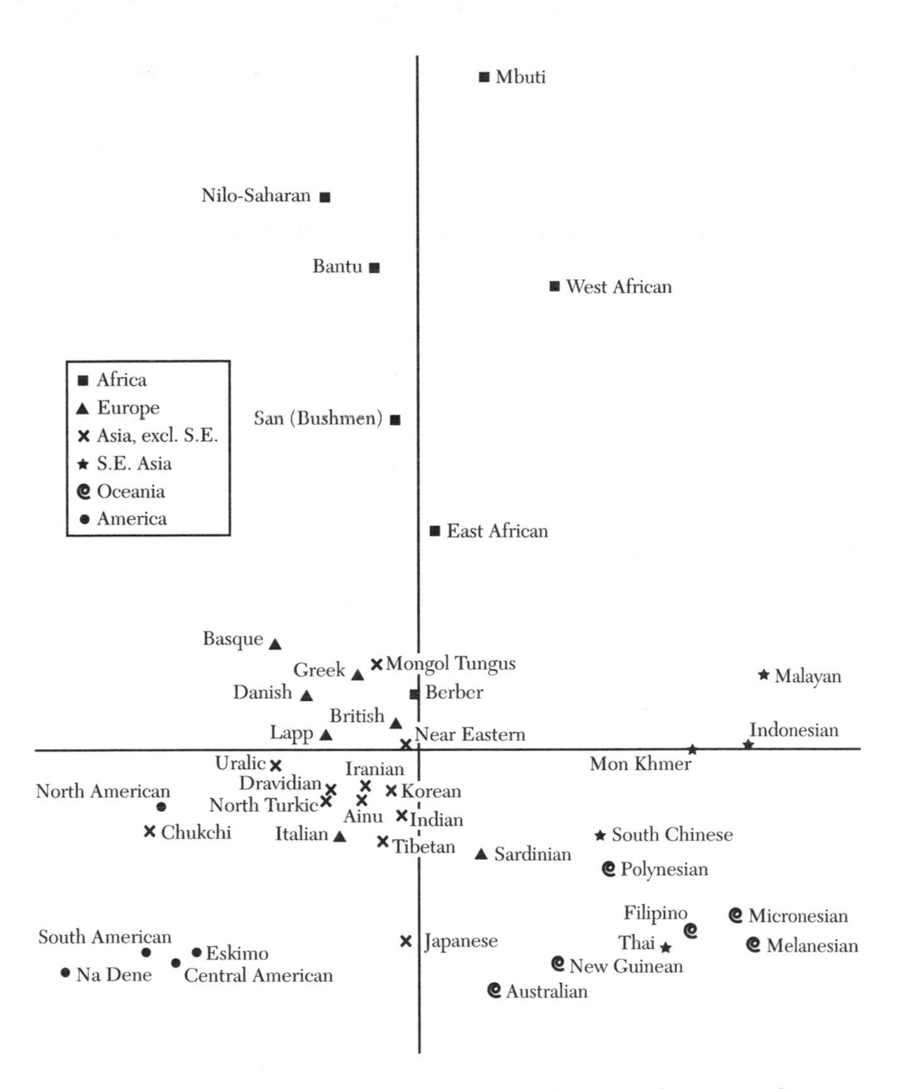

Figure 3. A synthetic view of 42 world populations, based on the genetic distances among them (see HGHG). The two-dimensional graph was built by multidimensional scaling, a variant of principal components analysis, and represents, with minimal loss of information, distances between pairs of objects (42 human populations in this case), calculated on the basis of many characters (110 genes). Populations are indicated by different symbols in order to distinguish the continents and subcontinents of origin. The only continents that are not differentiated by this analysis are Europe and Asia; an additional third dimension perpendicular to the first two would show that Europe is in a different plane compared to the rest of the world and differs less (on average) from Asia than the other two continents (Oceania and America), which were settled by migrants from Asia. (Computations by Dr. Eric Minch, while he was at Stanford University, using gene frequencies given in Cavalli-Sforza, Menozzi, and Piazza 1994)

many successive steps, every step bringing a small group of people, say from East Africa to North Africa, and others to southwestern Asia. Both groups grew in numbers and continued expanding. Those in southwestern Asia may have sent propagules north and east, and perhaps back toward East Africa. Further steps brought descendants from these groups to other parts of Asia, eventually reaching all the inhabitable world.

Figure 3 shows clearly that Asia was settled from Africa, while Oceania, Europe, and America were settled from Asia. For including Europe we need a little help from a third axis, which is not shown here. Illustrated in the lower left quadrant of the graph is the expansion from Asia to America, which took place from northeastern Asia via the Bering Strait (then not a sea tract but a land strip, Beringia). In the lower right quadrant, Southeast Asia appears as a propagule from mainland Asia; and, from it, Oceania was settled. We know that the settlement of Oceania happened by many successive expansions, the last of which is Polynesian, and is reasonably dated to the last 6,000 years.

Europe was also settled largely from Asia, as well as from Africa, as we have already mentioned, but the first two axes do not distinguish Europe and Asia. This is not too surprising, as there is no clear boundary between the two continents; the Urals do not constitute a formidable barrier. Eurasia can be viewed as a single continent. Nevertheless, the two continents are genetically distinct, as can be shown by adding a third axis to the graph. The third axis, perpendicular to the first two (not represented in our figure), shows Europe above Asia, on a plane higher than all other continents. But because of the geographic continuity with Asia, its genetic distance from Asia may be a little smaller than that from Oceania or America. The numerical information given in chapter 2 is not sufficient for making this statement, but the original data are much more informative, though inevitably more complicated.

In conclusion, history, given by trees, and geography, by principal components or multidimensional scaling, substantially agree. The simplest way to summarize both is to describe human evolution as having begun in Africa, where many groups became differentiated from one another over a fairly long time. African groups located closely to other continents expanded to them. The first and major expansion by African groups was from East Africa to Asia, probably via Suez and the Red Sea, perhaps continuing along the southern coast of southern Asia, but probably extending south also through the interior of Asia. Northeastern Asia was probably reached from the coast of Southeast Asia and from Central Asia. From Southeast Asia expansion to nearby New Guinea and Australia was natural and earlier than that to America. Expansion to Europe probably came from the east, west, and center, and was also relatively late. These statements are, for the time being, approximate and uncertain, because the genetic data collected so far are very limited. But it seems very likely that they can be made more precise in a fairly short time, because analysis of DNA, with methods already available or new ones in sight, provides many of these answers.

CHAPTER 4

Technological Revolutions and Gene Geography

Modern Human Expansions

Between 100,000 and 50,000 years ago, modern humans began to migrate out of Africa and adapt to new and diverse environments. Migration must have been a response to population growth and local overcrowding. Without growth, population density would fall in the region of origin, so these migrations should more accurately be called expansions.

Until 10,000 years ago, humans depended exclusively on hunting and gathering, a way of life that sharply limited population growth. We do not know precisely what the population size was 100,000 years ago, when people who were beginning to resemble contemporary humans inhabited Africa. Calculations based on the amount of genetic variation observed today suggest that the population would have been about 50,000 in the Paleolithic period, just before expansion out of Africa.

It is possible that the human species had nearly attained the point of population saturation in Africa when the expansion to the rest of the world began. When population density approaches saturation, humans, and probably all organisms, have a tendency to migrate to less populated areas. A very recent historical example is the great European migration to America and Australia in the last two centuries. The amount of territory available to Paleolithic Africans was both vast and generally accessible. The process, started in Africa, continued in each of the successive areas colonized.

High population density alone is probably not sufficient to initiate a geographic expansion, but it can stimulate cultural developments that allow or even encourage migration. The advent of sailing—even if primitive—may have aided some of the first expansions out of Africa. Boats were certainly needed to reach Australia 40,000 to 60,000 years ago. If they were invented earlier than that, they may even have been used to leave Africa for the southern Asian coast. It seems very likely that maritime navigation, however primitive, began in eastern or northeastern Africa. From the Red Sea the migration would have proceeded along the coast of southern and southeastern Asia, where it could then branch toward Oceania in the south and the Pacific rim as far as Beringia to the north (figure 4).

But I am convinced that another factor played a major role: the late Paleolithic expansion out of Africa was greatly served by the development of language. Our most distant human ancestors might have had some primitive linguistic ability, but the complexity characteristic of all contemporary languages probably wasn't attained until around 100,000 years ago. This formidable instrument of communication helped humans explore and establish small societies in distant lands, adapt to new ecological conditions, and rapidly absorb technological developments.

Be that as it may, demographic growth during the late Paleolithic was very slow. Agricultural development came at the juncture between Paleolithic and Neolithic, 10,000 years ago. Using ethnographic data collected from contemporary hunter-gatherers, we can approximate the global population density at this time. By extrapolation, we arrive at values between one and fifteen million inhabitants. Let's assume there were five million. This is a very slow rate of growth, from 50,000 people alive 100,000 years ago to five

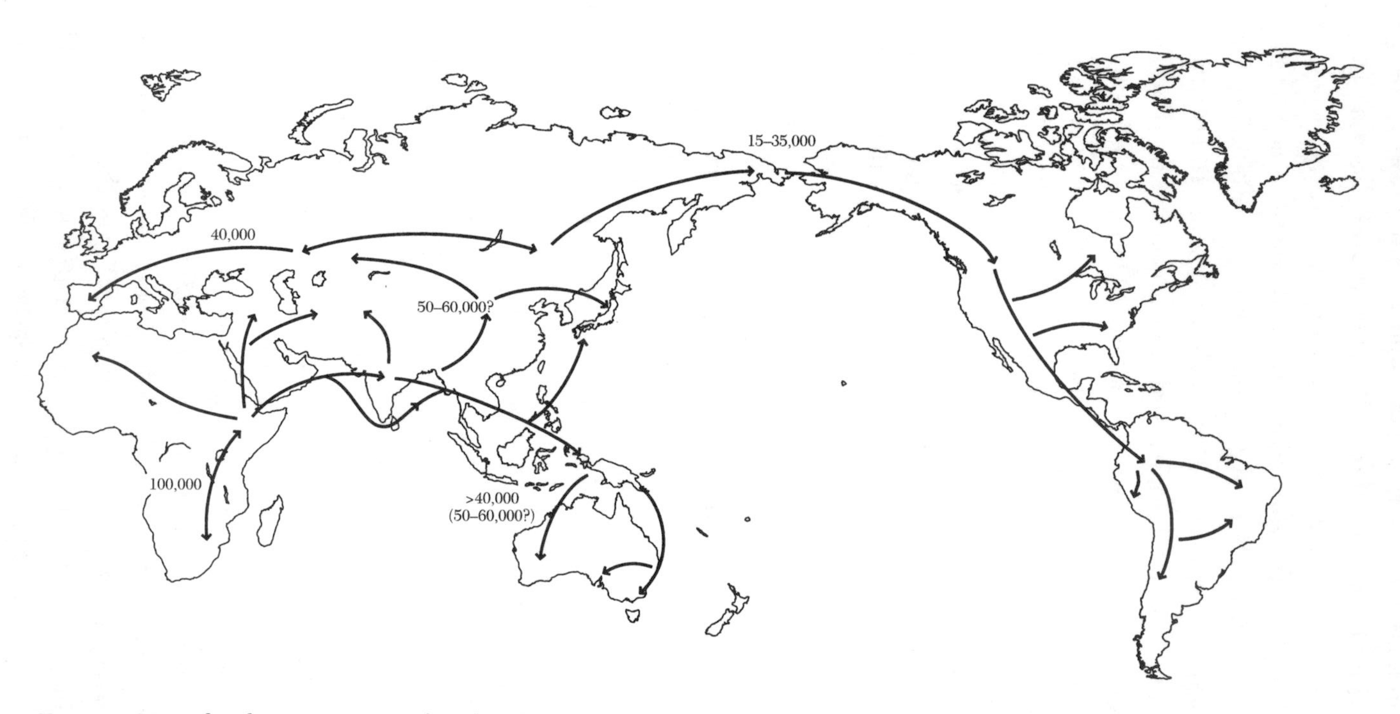

Figure 4. Map of earliest migrations of modern humans, beginning in Africa between 100,000 and 50,000 years ago, continuing into Asia and to the other continents, with approximate dates suggested by the archeological record.

million by the end of the Paleolithic. Higher growth rates followed agricultural innovation. It took 10,000 years to go from five million individuals to the present world population: a growth rate on average more than 14 times greater than during the Paleolithic.

In more recent times growth rates increased considerably: in the last century alone, the world population grew from 1.6 to almost 6 billion, nearly 250 times the average rate during the Paleolithic. We know that if the current pattern continues, the global population could reach a very dangerous point in the coming decades. Nature controls an excess of human births in three ways: epidemics, famine, and war. All of these brakes seem to be at work today: AIDS, an epidemic we still cannot control, is raging; extreme malnutrition affects more than a billion people; and an unprecedented number of civil and religious wars is shaking the world. So far, atomic bombs have not been used in these conflicts, but nothing should prevent us from worrying that a Russian scientist or engineer, reduced to unemployment and hunger, or a group of religious extremists working for a fundamentalist government, could place the human species at risk of a global Hiroshima.

The recent study of gene geography provides many examples of expansions—or diasporas, to use an ancient Greek word, which I shall use as a synonym for the numerical and geographic expansions of populations. Many significant diasporas occurred during both the Paleolithic and Neolithic periods. History records many that have happened in the last 5,000 years. Can we detect them in the geographic distribution of genes?

The small size of the human population during the Paleolithic favored the genetic differentiation of populations by genetic drift. Drift produces random variation in all genes. Therefore a great genetic distance between small populations is expected. Expansions across wide geographic areas encourage mixing between neighboring and distant populations, which leaves rather profound traces in the geography of genes. We can observe these migrations on geographic maps of genes even after several thousand years. When several successive migrations and expansions happen in the same area they begin to overlap and obscure each other, but it is often possible to recognize and distinguish each wave through a variety of statistical techniques—provided each has a different geographic origin.

Our analyses have shown that, in general, all the great expansions were due to important technological innovations—the discovery of new food sources, the development of new means of transportation, and the increase of military and political control are particularly powerful agents of expansion. The innovations responsible for the most significant expansions are those which produce local demographic growth and accompany populations as they move. The culture of cereal agriculture could be exported along with the cereals themselves. Wheat and barley, native to the Middle East, were domesticated there at the beginning of the Neolithic. At the same time almost all the domestic animals we are familiar with today were domesticated. The farming population grew and eventually spread to other fertile land where the cycle of demographic growth and expansion started all over. Expansion led anywhere the land was capable of supporting the growth of domesticated plants and animals, and would halt where the environment was not conducive to agriculture, as it did in the extreme north of Scandinavia and Russia, where it was too cold.

Fortunately, not all technological revolutions produce demographic growth and population expansions. One important period of population growth occurred in Europe during the second half of the Middle Ages. The increase was due to various agricultural innovations that reversed the economic decay that followed the destruction of the Western Roman Empire by barbarians. This demographic and economic expansion was restricted to Europe until the advent of transoceanic travel in the fifteenth century.

The First Agricultural Expansion

Many details about the Paleolithic expansion of modern humans may remain unknown to us forever, but more recent expansions are less mysterious. In collaboration with the archeologist Albert

Ammerman, I studied one such expansion stemming from the development of agriculture in the Middle East. This event is called the Neolithic transition because, at least in the Middle East, the shift from hunting and gathering to the cultivation of crops and animals was accompanied by new techniques of stone tool production from which the period takes its name. Somewhat later, the introduction of ceramics—probably invented elsewhere—prompted other developments that provide us with a very useful archeological marker. Such a marker can help track the spread of agriculture into Europe more reliably than Neolithic tools. The best marker, however, is the expanding presence of wheat or other crops that were unavailable before the expansion.

Population density was relatively high, considering the subsistence means of hunters and gatherers, at the end of the Paleolithic around 10,000 years ago, especially in the subtropical zones most favorable to human habitation. Around that time, a climatic change modified the fauna and flora, forcing humans to find new ways to gather food. Food production started supplementing hunting and gathering at about that time in at least three widely separated regions. The domestication of local plants and animals that already were a part of the human diet began in the Middle East, in China, and also in Mexico and the nearly adjacent northern Andean highlands of South America. Each of these areas developed unique practical strategies for food cultivation. In the Middle East, future farmers started growing several types of wheat and barley, and kept cows, pigs, goats, and sheep. Millet was farmed in northern China, rice and buffalo in the south. Pigs were raised almost everywhere. In the Americas there were corn, squash, beans, and many other plants, but few animals could be domesticated there. These changes occurred almost simultaneously, suggesting some external pressure such as global climate change paired with various individual changes in each of the three areas, such as the depletion of natural resources and demographic pressure. These last two factors might themselves actually result from, or be exacerbated by, climate change.

The oldest known ceramics were found in Japan. Approximately 12,000 years old, they mark an important moment in the area's history. Oddly enough, agriculture would not reach Japan for another 10,000 years, whereas in the Middle East ceramics appeared about 1,000 years after the local development of agriculture, and 3,000 years after they had appeared in Japan. It is difficult either to confirm or disprove that pottery came to the Middle East from Japan—indeed, it may well have been invented in both places independently. There was another earlier source of ceramic technology, which was closer to the Middle East, in the Sahara Desert. The Sahara—not in fact a desert at all then—supported in its mountainous areas a considerable population, as numerous paintings and carvings in the Tassili and Tibesti mountains prove. There is evidence in various oases of the Sahara that ceramics were being used at least a thousand years earlier than in the Middle East, but here also it is difficult to decide whether by independent invention or diffusion.

Population pressure was relieved, after a brief lag, by the increased resources provided by agriculture. Human populations rapidly multiply when living conditions are favorable. Even with only a primitive form of agriculture, populations will double with every new generation, and this continues to happen in many developing countries. At this rate it would take only a few centuries after adding farming to its hunting and gathering habits for a population to reach a new, higher saturation density.

If at all suitable, neighboring territory would be occupied by farmers in search of arable land. Primitive farmers lacked fertilizing techniques and had to leave fields fallow periodically, or search for completely new land. This was further impetus for expansion. The introduction of agriculture thus increased local population densities, and also favored geographic expansion to an extent governed by the local ecology.

Geographic expansion was easier from the Middle East than elsewhere, because wheat, barley, and domestic animals were well adapted to a vast surrounding area, including most of Europe, North Africa (which was not yet a desert), and western and southern Asia. In Mexico, corn and other crops spread north much more slowly, perhaps because of the difficulty of crossing a vast desert region, but they did diffuse southward. The spread of crops was more rapid in the Andes, where ecological diversity is greater. Much of the rest of tropical South America, excluding the Andes, was also slow in developing agriculture for ecological reasons. The variety of environments in and near China explains the rather different courses of its agriculture development. There were limits to the spread of agricultural innovation from these areas. A steppe bordered China in the north, there were deserts in the west, but Southeast Asia, including southern China, was suitable for rice cultivation

The Neolithic period began in the Middle East approximately 10,000 years ago. It started perhaps a little earlier than the agricultural revolution in Mexico or China. It lasted 5,000 years, until the coming of the Bronze Age. The agricultural economy spread from the Middle East in a northwestern direction toward Europe, but also eastward toward Iran, Pakistan, and India, and southwest to Egypt. Its agriculture was fairly complex, supporting a great variety of cereals and domestic animals. The expansion to Europe is better known than any other, because European archeology has been studied more intensively and for a longer time.

Domesticated cereals spread very regularly from their area of origin in the Middle East. Their dispersion in Europe is particularly well documented by archeologists. It took farming more than 4,000 years (starting 10,000 years ago, traveling one kilometer per year) to reach England through Anatolia (Turkey). It spread a little more rapidly along the Mediterranean coast, since it is easier to cover great distances by sea than by land. A map of the archeological dates by radiocarbon of the spread of wheat in Europe is given in figure 5 (see page 109).

Changes and adaptation to local climate were bound to happen. Expansion proceeded from Macedonia and Greece down the Mediterranean coast past southern Italy to the western Mediterranean. Obsidian tools are found early in the Aegean islands and prove that people knew how to build and use boats during the Neolithic. An intact Neolithic boat was in fact found in the Seine in France and another in Lake Bracciano in central Italy. Central Europe was settled by Neolithic people going up the Danube and down the Rhine and the other rivers of the European plain, where we have found a pottery with characteristic geometric decorations (so-called "linear" pottery).

The first farmers of the Middle East did not use ceramics, nor did the first agricultural colonists of Macedonia. But when pottery finally did arrive it traveled very fast, advancing into the rest of Europe along with agriculture, almost without exception. I said earlier that elsewhere—especially in Japan—ceramics were developed well before the advent of agriculture. There is confusion here in archeological terminology, because societies using ceramics are called "Neolithic" (a term applied to stone technology) in Europe and Japan. In Europe the term "Neolithic" is applied to agriculturalists who had not yet adopted ceramics (a delay of about a thousand years), while in Japan it is applied to people who used ceramics almost 10,000 years before they adopted agriculture.

I mentioned that it is difficult to exclude the possibility that pottery spread from Japan to the Middle East. It is possible that trade routes through Central Asia were established early. The Silk Road was so named because silk was carried on it from China to Europe in Roman times, and it was revived during the Middle Ages. There is earlier evidence of exchange between East and West by what might have been the same route. Huge cemeteries full of northern Europeans in the westernmost part of China, in Xinjiang Province, show that the path is quite ancient. This very dry desert region near the ancient basin of Tarim has thoroughly desiccated and preserved scores of bodies, especially those who, having died during the winter, were effectively freeze-dried. Some of these mummies have unmistakably blue eyes and blond hair. Their mtDNA confirms what can be observed with the naked eye. In addition, their equally well-preserved clothing seems to suggest a northern or central European origin. A fabric similar to modern Scottish tartan, which at that time was also made in Austria and Switzerland, was found on one body. Radiocarbon dates show that these people lived at least 3,800 years ago. They probably spoke the now extinct Indo-European language Tocharian, of which some writings in an ancient Indian script still survive. A fresco in China dating from the seventh century A.D. also shows elegantly dressed northern Europeans with blond and red hair. Victor Mair, an American orientalist, surmised from these recent discoveries that the Central Asian route connecting Asia and Europe may have opened very early, perhaps more than 4,000 years ago. It may well have been traveled at the very beginning of agricultural practice or earlier. Northern European peoples probably disappeared from this region as a result of Mongol expansions, although several genes of European origin persist in Xinjiang, the westernmost province of China. The Uighurs who live there are a population characterized by a great variety of complexions and show an approximately 3:1 ratio of Mongol to European mixture.

Demic Diffusion or Cultural Diffusion?

Albert Ammerman and I asked the following question: did migrating farmers bring agriculture along with them (a process we called "demic" diffusion), or was it only the knowledge and technology of agricultural production that spread ("cultural" diffusion)? Archeologists have shown little interest in this question for several reasons. First, it is very difficult to distinguish between these two possibilities using the archeological record alone. There is another difficulty, of a psychological nature. Archeologists working between the two world wars were trained to interpret every cultural event—from stylistic changes in axes and pottery to changes in burial practices during the copper and iron ages in central Europe—in terms of grand migrations and conquests. After the last world war, this approach was attacked, particularly by the English school of archeology. Researchers began to theorize that innovations could spread in densely populated regions as the result of well-developed commercial networks. This critique was important, but it eventually was carried to a dogmatic extreme. Before World War II, all cultural change was thought to result from massive migrations; afterward, migratory explanations were considered unacceptable. Only merchants traveled, carrying objects later recovered in diggings.

Archeology has shown that the spread of agriculture was very slow and was accompanied by a considerable increase in population density. By contrast, all purely cultural diffusions were very quick and rarely had demographic consequences. Ammerman and I asked, as critically as possible, whether the spread of agriculture in Europe was a cultural or demic process, that is, did farming or farmers spread? Its slow pace across the continent suggested a demic process, but would it be possible to predict the rate of demic expansion merely on the basis of the growth and migration rates of human populations? And, how would it compare with the observed rate of agricultural diffusion?

We were aided by a genetic theory developed by R. A. Fisher, which was easily applied to the ecological and demographic problems that concerned us. In the demographic formulation, the theory quantitatively predicted the rate of radial spread (starting from the center of an expansion) for a population that started searching for new territory when it was approaching the point of saturation. Without fertilization, which was unknown at the time, soil depletion is rapid and motivates people to move when overpopulation threatens. Naturally, migrants occupy the nearest unpopulated regions first. Nevertheless, there is a limit to the distance that primitive peasants could travel. Fisher's theory shows that a growing population spreads at an easily calculable rate that depends on two demographic variables: the population's growth rate and the migration rates. The archeological record showed that agriculture spread approximately one kilometer per year. It was a bit more rapid when people used boats or traveled along rivers or coastlines, and slower near physical barriers or areas of ecological change.

If the migration rate is low, a high population growth rate is needed to sustain an expansion of the observed speed. Conversely, if migration is high, the growth rate can be slow. The highest known reproduction rates, more than 3 percent per year, lead to a doubling of population size in less than a single generation. When met with this level of population growth, the migration rates observed for primitive farmers would lead to a rate of expansion as great or greater than that observed for the Neolithic expansion in Europe.

It is very difficult to measure the relevant growth rate from the archeological record, since the rate of change varies, diminishing continuously from the initial rate. The velocity for the most general growth curve, the logistic, is highest at the beginning and decreases to zero. But it is the initial rate—only briefly sustained—that matters to us. History shows that high growth rates are eminently possible when a population of farmers occupies a sparsely inhabited area. This was the case, for example, in the province of Quebec more than three centuries ago, where the original population included about 1,000 French women. They were recruited by Louis XIV as prospective brides to the men, mostly trappers and traders, who had earlier settled French Canada, and had no other chance of marrying French women. Louis XIV gave a dowry to each woman who agreed to marry under these conditions. These women, who often did not know their future husbands, were called "the King's daughters." The population grew at an explosive rate, almost as high as that of the first Dutch settlers in southern Africa, where the rate of growth (crudely measured, it must be admitted) was similar. Naturally, all of these peasants practiced a more refined form of agriculture than the Neolithic cultures, but their demographic behavior may be comparable.

Such rapid growth ensures that even very gradual migration would guarantee an expansion rate of one kilometer per year. We concluded that demographic data of population growth and migration are indeed compatible with the theory of demic diffusion of Neolithic cultivators.

But this hypothesis was not immediately welcomed by Anglo-American archeologists. Only recently has the situation begun to change. Colin Renfrew, Professor of Archeology at the University of Cambridge in England, enthusiastically endorsed the theory in a 1987 book, and in a 1989 *Scientific American* article. Several other archeologists have now accepted the theory we proposed in 1972. This is a prime example of how hard it is for new and revolutionary ideas to gain acceptance in the scientific world.

A Genetic Demonstration of the Demic Diffusion of Agriculture from the Middle East

Archeology can verify the occurrence of migration only in exceptional cases. Demographic studies of modern developing countries have helped convince us that the slow diffusion of agriculture is consistent with information about the growth and migration of primitive cultivators. Unfortunately this concordance can only suggest that the expansion might have been demic, but it cannot prove it with any certainty.

So we searched for new methods. One in particular turned out to be very satisfying: drawing synthetic geographic maps of genes.

A single gene cannot provide sufficiently clear and unambiguous results. Any gene is subject to the vagaries of chance, and maps illustrating a single gene frequency lend themselves to multiple, equally likely interpretations. As an example, let's discuss the geographic distribution of two well-known genes: the RH– gene in Europe, the frequency of which is highest in the Pyrenees Mountains and decreases all around, and the ABO blood group genes. Of these, the O form frequency reaches nearly 100 percent among American Natives, while B form frequency is maximal in East Asia and decreases toward Europe.

The RH– gene is a European allele—rare or entirely absent elsewhere. We can guess that the mutation from RH+ to RH– occurred in western Europe. Since we know that Europe was settled by modern humans about 40,000 years ago, the mutation probably occurred after that time, increased in frequency, and spread from its point of origin. Why did the frequency of RH– increase in the first place? It may have provided a selective advantage to its carriers, although it is difficult to imagine how or why, since the only selection we know of is that RH+ children of RH– mothers have a significant chance of suffering birth defects or even death. The risk exists for the second RH+ child of an RH– mother, and increases for subsequent RH+ children. Fetal damage is caused by the mother's antibodies to the RH+ gene raised during the first pregnancy with an RH+ child. We now know enough about this gene to minimize the risk to RH+ children; but it is still hard to imagine how the RH– gene could have increased in a predominantly RH+ population. It is worth noting that the RH+ gene would face a similar disadvantage in an RH– population. How can we explain the increase in western Europe of RH– genes?

Two hypotheses are possible: either natural selection, which may have favored RH– genes, is responsible for reasons unknown to us, or the RH– form has reached a high frequency through drift. As always, it is difficult to choose between these two standard alternatives. The drift theory is supported by the fact that the last glacial period began in Europe about 25,000 years ago, and reduced Europe's overall population, isolating western from eastern Europe, and most probably favoring genetic differentiation.

In chapter 2 we had to ask the same question about the ABO blood groups: did natural selection or drift cause the near disappearance of the A and B genes from the Americas, leaving the O gene at almost 100 percent frequency? The O blood type is rather frequent elsewhere, 50 percent on average, and the difference between 50 percent frequency in one population and 100 percent in another is hardly negligible. A possible explanation is that the trek over the Bering land bridge might have involved only a very small group of Siberian nomads, thus allowing drift (in the form of a "founders' effect") to erase any trace of the A and B genes. The A gene has been found in a part of northern Canada. It may have resulted from novel mutations or admixture with later American settlers, or other selection episodes. A further study of these genes at the DNA level in present-day populations and in earlier human remains might finally provide an answer.

On the other hand, natural selection could have eliminated individuals bearing the non-O blood groups, and a potential reason has been identified—syphilis, a disease that burst into Europe only after 1492. An event that helped spread the disease in Europe was a war against Spain fought by Charles VIII, king of France, near Naples, beginning in August 1494 and ending in February 1495 with the fall of that city. Naples was under Spanish control from then on, but the contagion spread from Spanish to French troops and the Italian population. The disease therefore acquired different names in different countries: Spanish, Neapolitan, French, and Gallic. The hypothesis of an American origin was suggested in the first, excellent scientific description of the disease, which gave the illness its name. This happened, according to the customs of the time, in the Latin poem "Syphilis sive Morbus Gallicus" (Syphilis or the Gallic disease), written by Girolamo Fracastoro in 1530. In the poem, a young American shepherd named Syphilis is unfaithful to the sun god and plagued by ghastly syphilitic ulcers as punishment. But the god forgives him and teaches him a treatment involving an American plant and mercury. In another work, "De contagione et contagiosis morbis" (Of contagion and contagious diseases), 1546, Fracastoro interprets infectious diseases including syphilis, leprosy, tuberculosis, typhus, and so on, in a remarkably modern way. Fracastoro's extraordinary intuition in all these matters leads me to believe that his theory on the American origin of syphilis, and its transportation to Europe by Christopher Columbus's sailors, is also correct. The theory is given further credence by our current knowledge that type O individuals under treatment for syphilis recover more rapidly (from an immunological viewpoint) than those belonging to other blood groups.

As mentioned before, it is generally hard to explain the geographic map of a single gene. In the case of ABO I believe both hypotheses of drift and selection are correct. The map of the RH– gene is compatible with the diffusion of Middle Eastern farmers, if we agree that the Neolithic cultivators were predominantly RH+, like the rest of the world, and that in the Paleolithic period, western Europeans were mostly, or all, RH–. However, many other explanations are possible.

Fortunately, a number of genes other than RH agree with this interpretation. Only those genes that had frequencies different in the population of the Middle East than in the tribes living in Europe before the dispersal of Neolithic groups can provide us with useful information. We don't know in advance which genes displayed such differences between the two regions, but we would guess that those that show a significant gradient from the area of origin to the area of ultimate arrival today probably differed in the two regions 10,000 years ago as well.

Before the development of agriculture, population sizes were small, and much genetic drift was expected, producing widely differing gene frequencies from one area to another. Because the Neolithic populations that came later produced food much more abundantly than earlier populations, they could reach much higher population densities than Paleolithic populations. They would expand to neighboring areas and their genes would not be completely diluted by their migration into Europe and their subsequent mixing with native inhabitants. But we would observe a progressive dilution of genes emanating from the Middle East in their passage across Europe.

Unlike selection, migration affects all genes equally; as a result, we can reconstruct ancient migrations in the course of drawing geographic maps summarizing all available gene frequency information. The more genes we study, the more reliable our results. Ammerman and I had data for only 39 genes when we first started in 1978. Having repeated our analysis with 95 genes now, the results are extremely similar, but more precise. It seems likely that Europe has experienced many major migrations at different times, the traces of which have been superimposed upon each other. Europe is the most studied continent from a number of perspectives, including genetics and archeology. Can we disentangle all these migrations?

Principal components analysis, which we discussed earlier, helped us do just that. Principal components are unique quantities that essentially summarize most of the information contained within the frequencies of many genes or, in general, of many variables. Each component may successfully isolate one from the other different factors influencing variation of the gene frequencies at geographic points, and many of these may be different migrations and expansions.

In order to calculate principal components (PCs), it was first necessary to create geographic maps for each gene studied throughout Europe and the Middle East. We drew maps for the 39 genes for which we had sufficiently detailed data and then we calculated the PCs with the help of a computer. Finally we'd draw up the geographic maps for each component. The total variation was thus decomposed into its "components," as the PCs are aptly called (the word "principal" refers to the fact that there would be many more components, but we have chosen the most important ones). We could also calculate the fraction of the total variation summarized by each component. The most important components are those that explain the largest fraction of the variation. The method works sequentially: it first calculates a component that can represent all the gene frequencies with just one value. This value, which is called the first principal component, is a sum of the gene frequencies observed at a specific point on the geographic map. But each frequency has been previously multiplied by a value that is different for every gene frequency, and serves as a "weight." The weight of each gene frequency is calculated by a mathematical procedure that assigns a relatively large value if the gene frequency is important in determining overall genetic variation, and a small one if it is not. One could describe a PC as a "weighted mean," in which all gene frequencies are averaged but each particular frequency is given more or less weight as suggested by a precise method of computation.

After the first PC has been calculated, it is eliminated from the data, and we proceed to calculate a new principal component from the variation of the gene frequencies that remain available after subtracting the first PC. This is the second PC, and the method continues to calculate all successive ones. Each PC is independent from all others, and uses a different set of "weights" by which each gene frequency is multiplied before adding them up to generate the value of the principal component itself.

The total number of components possible is the number of genes minus one. Only the first few are really meaningful. The method also calculates the fraction of total genetic variation accounted for by each component, and this fraction decreases with the order of components. Hence the first principal component is the most important. In our first attempt, we calculated only the first three components, and together they explained about half of the total genetic variation.

To our great astonishment, we saw that the first principal component of the geographic map in Europe, as illustrated in figure 6, perfectly matched the map plotting the arrival dates of cereals in Europe according to radiocarbon estimates (figure 5). Paolo Menozzi, collaborating with Alberto Piazza and me, performed the plotting of the principal component maps. He never thought the result would be so seemingly accurate, and when he saw the maps his pleasure was as great as his surprise. The correlation between the archeological and genetic maps is obvious, and was confirmed

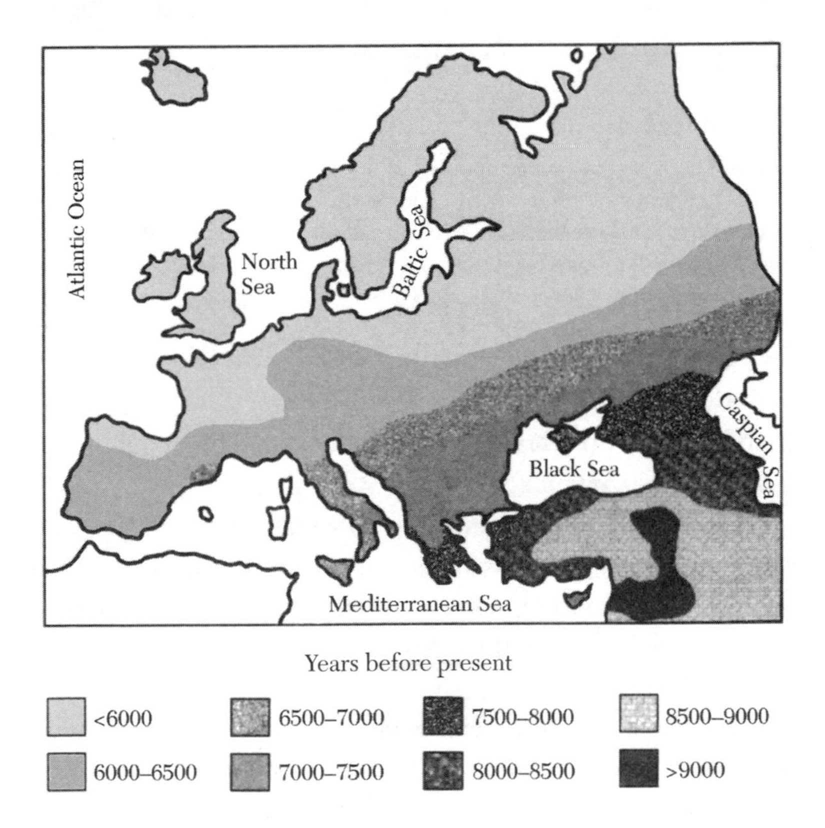

Figure 5. The spread of agriculture—specifically, the arrival of wheat from the Middle East to the various parts of Europe, from 9,500 years to 5,000 years ago. (Redrawn from a map prepared by Ammerman and Cavalli-Sforza 1984)

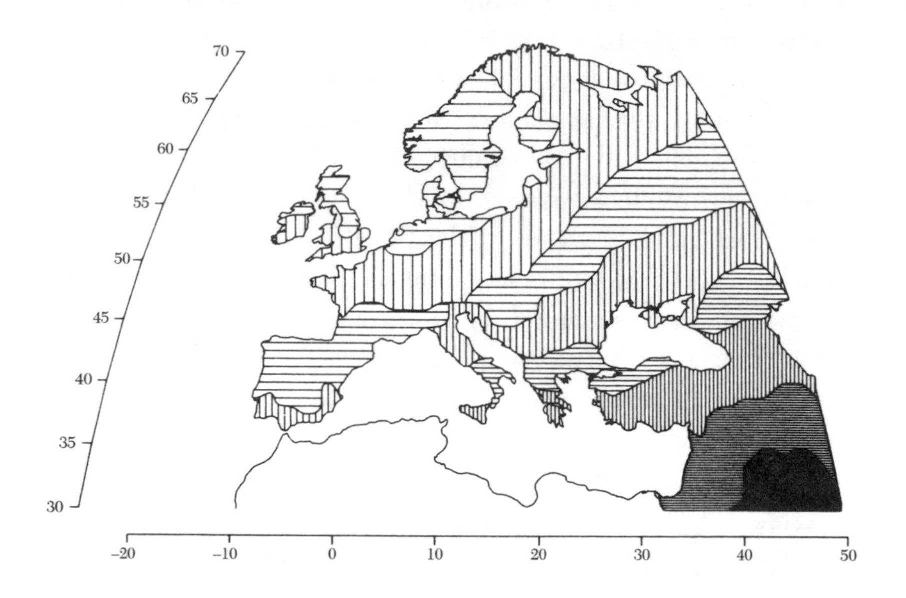

Figure 6. The first principal component of 95 genes in Europe. Its extraordinary similarity to figure 5 representing agricultural diffusion indicates that—by the simplest interpretation—there was an expansion of farmers from the Middle East into Europe, who, in the course of expansion, mixed with local hunter-gatherers, who had different gene frequencies. (Figures 6 through 10 are taken, with permission, from *The History and Geography of Human Genes,* by L. Cavalli-Sforza, P. Menozzi, and A. Piazza, published by Princeton University Press in 1994.)

with independent methods by Robert Sokal and his collaborators at the State University of New York at Stony Brook.

So far, we have not explained the meaning of the bands of varying density in the principal components map. In the maps of single gene frequencies each band represents an arbitrarily chosen range of gene frequencies, for instance, the range from 10 to 20 percent of a given gene. But the principal components are calculated using the average gene frequencies of many genes, weighted by coefficients calculated by methods too complicated to explain here. What scale do we use to represent these PC maps? The original values are centered on the average value of each component, which is set equal to zero. They extend from the average in a negative and positive direction, with a scale dictated by a frequently applied statistical convention that is essentially arbitrary (for readers familiar with elementary statistics, the components are expressed in units of standard deviation). Discerning readers may be disappointed that I do not precisely specify the scale of measurement, but it is not easy to explain in just a few words. I encountered the problem at its worst in 1994 when *The New York Times* published an article on this research and wanted to give an explanation of the scale of the principal components. The approach they took, without consulting me, was incorrect. In a legend to figure 6, they wrote that one end of the scale was "less similar" and the other end was "more similar." But they did not try to answer the question inevitably raised by this explanation: similar to what? I could perhaps have told them: more similar to the genetic types present at the source of the expansion. But that would be only an approximation, because the extremes of the scale are difficult to define with precision. The central value of the scale corresponds to the mean genetic type of each component in the geographic region being studied, and the two directions of a component—negative and positive—express the difference from the mean relative to the two poles. In practice, the extreme value of one of the two poles represents the center of an expansion, appearing at the center of radiating bands, and the opposite extreme indicates the regions genetically most different from the originators of the expansion.

Subsequently computer simulations by Sabina Rendine et al. (1986) have shown that we can effectively separate independent expansions by this method, especially if each expansion has a substantially different geographic origin, and if local populations experience only partial replacement by a foreign and genetically distinct expanding population.

It is critical that the expanding population have a demographic advantage over the recipient population with which it mixes—if not immediately, certainly by the end of the process. Neolithic farmers undoubtedly had higher population densities than the Paleolithic populations, and for that reason the Neolithic transition dominates the genetic backdrop of Europe even today. Recently, in northern Germany, near coal mining operations around Cologne, German archeologists have been able to record a big peasant expansion in the spread of "linear" pottery, the name given to the ancient culture of early Neolithic farmers in central Europe. Archeological excavations showed that Neolithic population densities were elevated, as one would expect. Furthermore, our computer simulations demonstrate that genetic gradients resulting from progressive mixing are fairly stable over time and could have persisted without much change over the fifty centuries which have elapsed since the end of the Neolithic.

The use of principal components analysis may seem unduly complicated to persons who dislike mathematics, or a simplistic treatment to those who know its mathematical background, called "spectral analysis of matrices." But, as we tried to explain in our first paper on this analysis, and showed with further simulations, the method is very efficient for disentangling superimposed migrations. The multiplication of all gene frequencies by appropriate weights and their sum is what mathematicians call "a linear analysis." Principal components are statistically independent from one another, and so can isolate independent expansions. Migration transforms gene frequencies "linearly," and migrations arising at different times from different origins are most likely to be independent, that is, "uncorrelated." Perhaps this explanation seems complex, but before the method is too easily dismissed, one should note that, from the point of view of evolutionary theory, it is clear that principal components are the most satisfactory method for isolating independent migrations.

We should point out that the populations that may most closely resemble the Paleolithic and Mesolithic Europeans before the arrival of Neolithic populations are Basques. They speak a language completely unlike that of any other Europeans. The work of Michael Angelo Etcheverry, Arthur Mourant, and Jacques Ruffié on the RH gene had already suggested a proto-European origin for the Basques on the basis of genetic evidence. Our study is in perfect agreement with this proposition and indicates that the Basques are likely to have descended directly from Paleolithic and then Mesolithic populations living in the southwest of France and northern Spain before the arrival of Neolithic peoples. Like all other ancient populations, the Basques have gradually mixed with their new neighbors. They are not a purely Paleolithic people in this sense, but thanks to partial endogamy (marrying mostly within their ethnic group—aided in part by their difficult and unique language) they have maintained some genetic distinctiveness from the neighboring populations, which must at least partially reflect their original genetic composition.

Very confident support for our conclusions came more recently from research with Y chromosome markers. An unmistakably strong east–west diffusion from the Middle East to Europe was proved in 1997 by Ornella Semino and other Pavian population geneticists directed by Silvana Santachiara Benerecetti, using two major markers. Their results initially conflicted with the work done on mitochondrial DNA by Brian Sykes's group at Oxford, but expanding the number of individuals used altered their conclusions. With Giuseppe Passarino of Cosenza, Peter Underhill of Stanford, and others in my laboratory, Semino extended the Y chromosome research to seven new markers on 1,000 Europeans. These unpublished results dramatically confirm the expansion of farmers from the Middle East, and build on conclusions drawn from the second and third principal components. The results suggest postglacial expansions from glacial refugia in southern France and eastern Europe, and provide new information about more recent expansions from central and eastern Europe.

Other Principal Components of the European Genetic Landscape

The other components after the first, which as we have seen is connected with the expansion of agriculture from the Middle East, have revealed other expansions and phenomena of biological and historical interest.

The second component (figure 7) shows a north–south cline of variation, suggesting a correlation with climate. Another phenomenon that is superficially different, the distribution of languages, is also related to both the genetic and the climatic gradient. The languages spoken throughout most of northeastern Europe belong to the Uralic

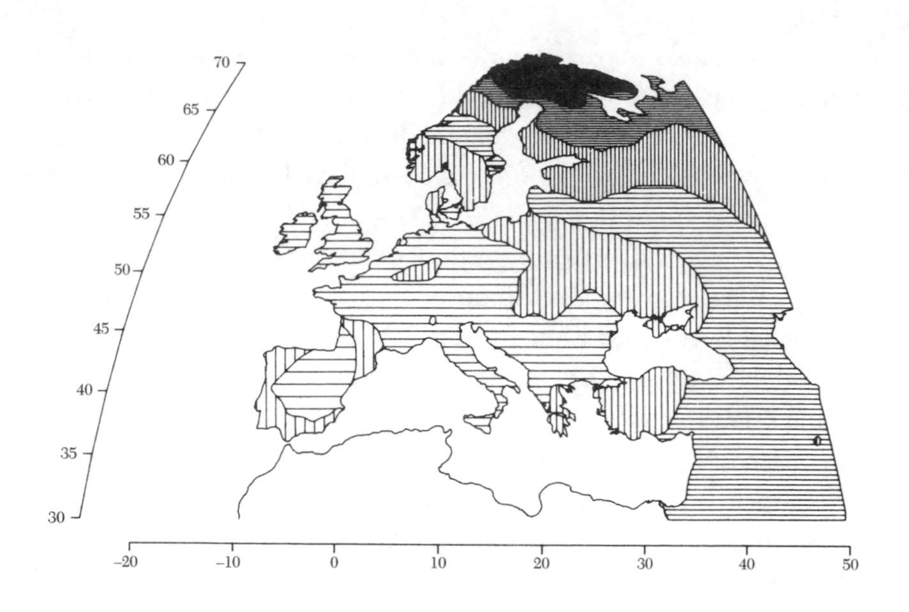

Figure 7. The second principal component of 95 genes in Europe. There seem to be two major streams of gene diffusion, most probably due to two expansions (one centered in the northeast and the other in the southwest of Europe) that took place after the end of the last glaciation.

family. They are very different from the languages spoken in the rest of the continent, which belong to the Indo-European family. Uralic languages are spoken mostly to the east of the Ural Mountains, but a number of them are found to the west. For example, the Saami (unfortunately most people are familiar with their other name, Lapps, which is derogatory) and Finnish languages belong to the western Uralic linguistic subfamily. The Indo-European family is composed of languages spoken from Spain and England in the west to Iran and India in the east. There are a few interruptions: in the Pyrenees where Basque survives; in Hungary (Hungarian is related to Finnish) and southeast of Finland (e.g., Estonian, Karelian); and in Turkey, where the language belongs to a totally different linguistic family (Altaic). We know that Latin was the administrative language in the ancient Roman province of Pannonia, which corresponds roughly to modern Hungary, but Pannonia was invaded by the Uralic-speaking Magyars at the end of the ninth century A.D. They imposed their language on the province, a frequent outcome of conquests.

Southwestern Europe was occupied first by people accustomed to the warmer climate. Does the second principal component of gene frequencies indicate genetic changes due to adaptations to the colder northern latitudes, as the correlation with latitude suggests, or due to the arrival of Uralic-speaking populations from western Siberia? It is possible that both explanations are correct and represent the same phenomenon from two totally different perspectives, biological and linguistic.

Another explanation has been recently suggested by Antonio Torroni, on the basis of a mitochondrial DNA study of European populations. He postulates that the second PC illustrates an expansion from southwestern Europe after the end of the glaciation, around 13,000 years ago. It is perfectly possible that this explanation also is correct. The expansion from the northeast has its center in the area with the darkest bands, Saami country; that from the southwest in the lightest area, in Basque country. Both poles of the second PC show a pattern rather similar to the expected genetic picture of an expansion, and it may well be that the second PC arises from two expansions, starting from the two extreme corners of Europe and proceeding toward the middle. That from the northeast was probably later; both involved hunter-gatherers, at least at the beginning, and were slow. One consideration gives some support to this idea. An expansion should proceed from a center toward the periphery and, if unimpeded in all directions, would generate a pattern resembling the circular waves generated by a stone thrown in a pond. Geographic irregularities seldom permit this; in the case of the first PC we do observe, however, an expansion into a circular sector of almost 90 degrees, resembling a fan with the pivot in the Middle East. But the map of the second PC seems to have one center of origin in the Basque region, opening like a fan toward the east and northeast; and another in the northeast, generating a fan toward the southwest. It is possible that there were two expansions exactly opposite to each other. The relative contributions and timings are hard to evaluate, but it is likely that the one starting in the Basque region is earlier.

Some further aspects of the history of Uralic language speakers, who live mostly in the extreme northeast of Europe and northwest of Asia, are of interest. They probably had sufficient time to adapt to the cold by biological or cultural adaptation, or, more likely, both. The Urals would not have represented a great barrier to travel, but nevertheless, the Saami are the only population on the western side who show some genetic continuity with populations living further east. These people, accustomed to the snow, who apparently knew how to make and use skis at least 2,000 years ago, could rapidly traverse the frozen plains.

The Saami are genetically European, but they also have affinities with non-Europeans, probably as a result of their trans-Uralic origins. Their European genetic resemblance suggests that their Uralic origins are partly masked by admixture with North Europeans, or vice versa. In any case, the European genetic element predominates. Other European Uralic speakers (e.g., Finns and Estonians) appear almost entirely European genetically. As to Hungarians, about 12 percent of their genes have a Uralic origin. Also in maps of the second principal component, we see in the lines representing equal gene frequencies a deviation to include Hungary, demonstrating a slight connection with northern populations, especially the Saami. The Finns, by contrast, show almost no trace of genetic admixture with Uralic populations, but there is another explanation for this. Finnish scientists have shown that their population displays a very unusual array of genetic diseases: some genetic problems, very rare or entirely unknown elsewhere, are sometimes very frequent in Finland, and vice versa. The genetic explanation of this observation is very simply extreme genetic drift. It is a common phenomenon in all populations that originate from a small number of founders, or have suffered at some later stage a severe reduction in population size: their pattern of genetic diseases is severely altered. The reason is the abnormal statistical fluctuations observed in small populations.

The probable scenario is the following. The very small group that gave origin to modern Finns entered the plains of Finland 2,000 years ago, from the south or east. A Saami population already inhabited this area, and eventually retreated to the north. Contact of Finns and Saami was enough for the Finnish immigrants to learn their language, even though substantial genetic mixing did not occur. Especially if several small groups of settlers speaking different languages entered the area, they all had to learn the language, or the local dialect, of the only people who knew how to survive and get around in Finland's maze of lakes. A similar situation is occurring in Mozambique, where a variety of local Bantu languages are spoken, but Portuguese, the language of their colonizers, is used for communication between tribes.

The third principal component is extremely interesting. Figure 8 is a little different from those reproduced in our other recent publications, because we have been able to add new data collected by Dr. I. S. Nasidze from the critical region encompassing the Caucasus (Piazza et al. 1995). The overall appearance of this map and its predecessors may at first appear very similar, but this one is more

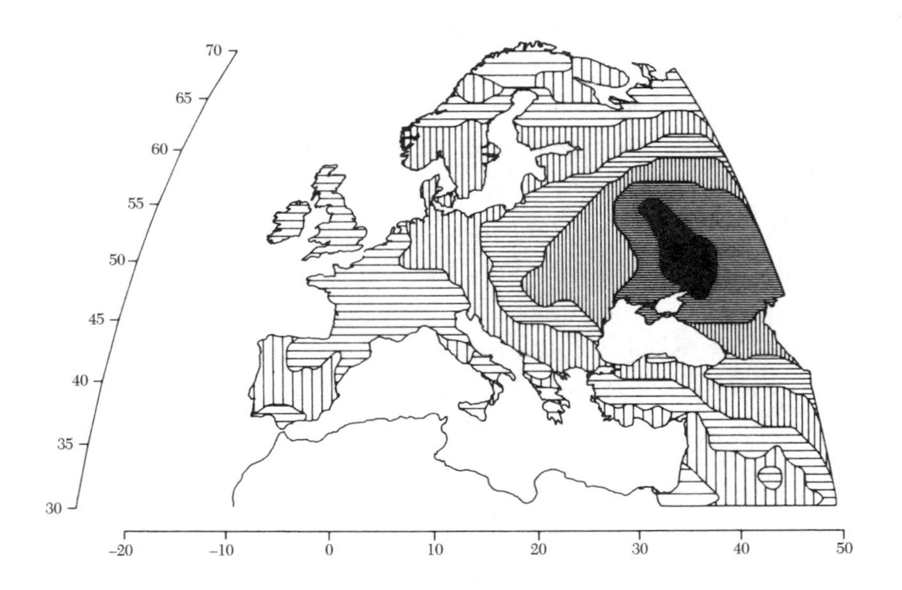

Figure 8. The third principal component of 95 genes in Europe reveals an expansion from a region north of the Black Sea (as the archeologist M. Gimbutas claimed) by pastoral nomads who domesticated the horse in the steppe. According to Gimbutas, they were responsible for building churn tombs in the region of origin, and for spreading Indo-European languages.

robust statistically. It shows an expansion originating in an area north of the Caucasus and the Black and Caspian Seas, which the archeologist Marjia Gimbutas had already proposed as the homeland of Indo-European speakers.

We shall discuss the evolution of languages in the next chapter. Suffice it to say here that much discussion has centered on the geographic origins of the Indo-European languages, with suggestions spanning from central Europe to Central Asia. Marjia Gimbutas has suggested that the Indo-European languages spread from a region north of the Caucasus and south of the Urals, where numerous tombs called kurgan have been found. These tombs were filled with sculptures, precious metals, bronze weapons, and the skeletons of both warriors and horses. Ecologically, the area belongs to the Eurasiatic steppe, which extends almost without interruption from Romania to Manchuria. Horses were common in the area, and the archeologist David Anthony has recently shown that they were probably domesticated in the vicinity of this Kurgan culture, where chariots and bronze weapons were made more than 5,000 years ago. Without written documents, it is very difficult for archeologists to say what language was spoken in this region at the time.

Another archeologist, Colin Renfrew, has offered a different hypothesis: he believes Indo-European languages originated from Anatolia in modern day Turkey. The first farmers of this area would have spoken a proto-Indo-European language, and would have spread it across Europe. Renfrew based his hypothesis on the belief that agriculture was spread by farmers, not culturally, and farmers would have had to bring their own language with them. This hypothesis has received less philological support than Gimbutas's. But as we shall see later, the two theories are not completely contradictory.

The people of the Kurgan culture were pastoral nomads who domesticated horses in the steppes where agriculture was not very productive. The horse provided milk, meat, transport, and as the Kurgan people would discover later, military power. But these nomads might originally have descended from agriculturalists of the Middle East or Anatolia, who probably arrived on the steppes through Macedonia and Romania, and may have spoken a pre–proto-Indo-European spoken in Anatolia at the beginning of agricultural development, around 9,000 to 10,000 years ago. Thus the language(s) common to Anatolia 9,000 to 10,000 years ago were earlier forms of Indo-European that spread locally to the Balkans and to the steppe. The languages that developed from this early Indo-European in the Kurgan region were later spread by pastoral nomads to most of Europe, beginning 3,000 to 4,000 years later.

The percent of the variation explained by lower principal components and their corresponding importance steadily diminish by definition. Nevertheless the fourth and fifth components are still statistically reliable in Europe and can be simply explained. The fourth (figure 9) shows an expansion from Greece toward southern Italy, called in Latin Magna Graecia (Greater Greece) because southern Italy became more important and populated than Greece itself. Greek expansion also included Macedonia and western Turkey. We know that the Aegean Islands had a long history even before historical Greece, and we admire the excellent art of these

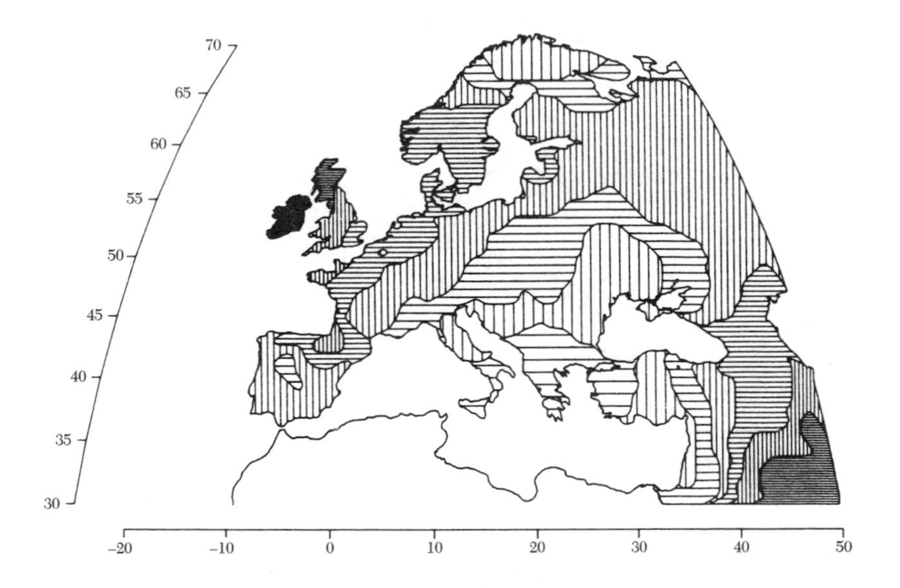

Figure 9. The fourth principal component of European genes seems to indicate the Greek colonization of the first millennium B.C.

ancient islanders. Homer relates only the destruction of Troy, which occurred around 1300 B.C., but the city flourished long before. Cretan civilization had a script, linear A, before 1400 B.C. Linear A was probably not a form of Greek; the first examples of written Greek survive in a later Cretan script similar to linear A, called linear B. The Greeks began a systematic colonization of southern Italy around 800 B.C.

The fifth principal component (figure 10) shows a pole in the easily identified Basque homeland. This component repeats on a smaller scale the lower expansion in the second PC. Today, the Basque language and culture survive in southwestern France and northern Spain, in the western Pyrenees. Historical information from Roman times, place names (toponymy), and genetics all confirm that the Basques once inhabited a much larger territory than today. The area in which Basque is still spoken has sharply contracted, especially in France where, under pressure favoring the French language, Basque is spoken by only about 12,000 people.

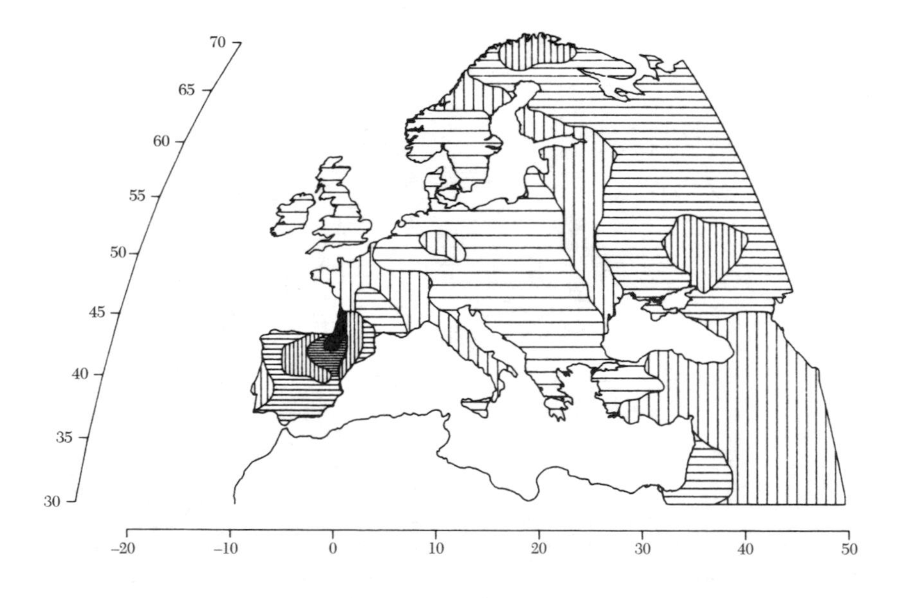

Figure 10. The fifth principal component corresponds to the area occupied by speakers of the Basque language. (Maps of principal components in figures 6–10 taken from Cavalli-Sforza, Menozzi, and Piazza 1994)

It's spoken by many more people in Spain. During the last Paleolithic period the Basque region extended over almost the entire area where ancient cave paintings have been found. There are some cues that Basque descends from a language spoken 35,000 to 40,000 years ago, during the first occupation of France by modern humans, who most probably came from the southwest, but possibly from the east as well. The artists of these caves would have spoken a language of the first, preagricultural Europeans, from which modern Basque is derived.

Population Expansions Outside of Europe

We have seen that agriculture spread in many directions from the Middle East toward other independent centers of agricultural origin. The eastern expansion toward Iran and India is clearly visible in the genetic maps of Asia. The same wave of expansion also headed toward Arabia and North Africa. However, as in many of the regions that would later become deserts, few of the original populations have survived. The replacement of Neolithic populations with more modern ones occurred most extensively in what is now the Sahara Desert. We find significant areas of admixture between Whites (Caucasoids) and Blacks in Africa: throughout most of the Sahara where Whites have crossed both the Suez and the Mediterranean, and in East Africa as a result of late Arab contact, which is well documented historically. Ancient cave paintings in the Sahara make it clear that the earliest Saharan populations were Black—although possibly mixed with Caucasoids—up to about 5,000 years ago. The most beautiful fresco in the Tassili, near a locality known as Jabbaren, shows two young attractive women. Today they are usually called the young Peulh (or Fulani) after the black population that now lives in the Sahel—a semi-desert strip south of the Sahara. The Peulh are typically nomadic pastoralists, who live off their cattle herds as their ancestors did before them. The paintings found in the mountains of the Sahara also depict many cows, most of which had been domesticated.

The Berber populations nearer the Mediterranean coast were probably Caucasoids. There is little doubt that they came from the Middle East, and they have occupied the region since the Neolithic or even earlier. Experienced sailors like other Neolithic peoples, they colonized the Canary Islands. When the Spaniards conquered these islands in the fifteenth century, they found a distinct population with some blond-haired and blue-eyed people—traits that are still evident among some Berbers in Morocco. They spoke Guanche, an Afroasiatic Berber language. By the time the Spanish arrived, they had lost the ability to sail.

For the most part, the Berbers were forced into the interior or mountain refuges by the arrival of Arabs beginning in the seventh century A.D. The Tuaregs, the dominant population of the Sahara itself, also speak a Berber language. They are genetically very similar to the Beja, another group of desert pastoralists who live along the Red Sea coast of Sudan, in the extreme east of the Sahara.

Today, a few groups continue to live in the mountains of the Sahara, and are generally much darker than the Berbers, the Tuaregs, and the Beja: the Teda live in the Tibesti mountains of Chad, the Daza in the Ennedi, and the Nubans in the Kordofan hills in Sudan. It would be most interesting to compare all these groups with the more recent and powerful molecular techniques, if sufficient material from these remote populations is available. The darker groups may be more direct descendants of the Saharan potters whose work predates the ceramic industry in the Middle East. It is reasonable to hypothesize that in the last 5,000 or 6,000 years, White populations arrived in the Sahara from the north or the east and either mixed with or partially replaced the area's first inhabitants, who were black.

The Sahara began to transform about 3,000 years ago into the harsh desert it is today. Horses were replaced with more drought tolerant camels imported from Asia, and peasant populations were forced south.

We are not sure if cows were domesticated in North Africa before they were in the Middle East, but several lines of archeological and genetic evidence favor this interpretation. Early Saharan rock paintings show that bovines were numerous. Herders who pushed south found that cows cannot survive at the periphery of tropical forests of West Africa and central Africa, where the tsetse fly transmits a bovine version of sleeping sickness. Only the savannas south of the Sahara would support the herds of the Saharan pastoralists. These herders had a characteristic body morphology: they were tall, thin, and had long arms. It is possible that this morphology—called "elongated" by the French anthropologist Jean Hiernaux—is an adaptation to life in an extremely hot and dry environment. People there often speak Nilo-Saharan (Nilotic) languages.

The farmers who left the increasingly dry Sahara, 3,000 to 4,000 years ago, found south of the desert conditions favorable for growing local domesticates like sorghum, millet, and other cereals, as well as cattle, sheep, and goats. In Mali and Burkina Faso, an early demographic expansion seems to have resulted from the development of agriculture, but archeological information is lacking from this important region. It is the study of genetic variation that gives us the impression that a demographic expansion occurred there, and I hope that archeologists will take notice. Further south, however, a much more radical solution was needed, since northern domesticates could not be grown in the tropics. Completely new plants, mostly roots and tubers from the local forests, were domesticated. However, completely satisfactory solutions for tropical agriculture were not found in Africa. Only much later, two very similar roots—both called manioc or cassava—that had been domesticated millennia before in the South American forest were introduced to Africa, possibly by missionaries in the eighteenth century, and gained immediate success throughout the African forest. They are now the most popular food and major source of calories in a very wide range of tropical Africa.

Western Africa witnessed the growth and spread of several groups farming local plants and cereals. The strongest evidence for these expansions from Senegal, Mali, Burkina Faso, and especially Nigeria and Cameroon comes from linguistics. The most dramatic demic expansion started from the vicinity of Cameroon about 3,000 years ago, or even earlier, in late Neolithic times, and was helped along by the use of iron beginning around 500 B.C. This has been called the Bantu expansion after the languages spoken by its protagonists, which comprise the most recent but also most successful branch of the major linguistic family of Africa, called Niger-Kordofanian. The expansion resulted in the rapid occupation of central and southern Africa by Bantu speakers. They were about to reach the Cape of Good Hope when the Dutch built a colony to supply their ships headed for India—with well-known consequences. As was already noticed by Jean Hiernaux, genetics clearly show that the Bantu are relatively homogeneous and distinct from other West Africans—their closest relatives. They have mixed with Nilotic speakers in East Africa, and with Khoisan speakers in the south. Hiernaux correctly deduced that a demic expansion had occurred. It lasted a little more than 3,000 years and was about 1.5 times more rapid than the European Neolithic expansion. Indeed, in the second phase of their spread, these people used a slightly more advanced technology, iron, than was available to the Neolithic Europeans, who were still in the stone age.

In China, we find independent but nearly simultaneous agricultural developments in the north, east and south. In the northern Xian Province, later the center of the Qin and Han Dynasties (from about 220 B.C.), millet and pigs were cultivated with tremendous success. In southern China, rice and buffalo were farmed. There were two or three important agricultural centers of origin in the south. One of them included Taiwan, which was attached to the mainland until quite late, and would later function as the source of massive migrations first toward the Philippines, then Melanesia and Polynesia.

The two parts of China were very different during the Paleolithic, and this dichotomy is still visible in the modern residents. Genetically, northern Chinese resemble Manchurians, Koreans, and Japanese. Southern Chinese are more like Southeast Asians. China has been unified for more than 2,000 years, and while there has been internal movement, it has remained genetically and culturally divided. The north and south are two worlds; although bound by a common language and political base, they have maintained some of their former divisions.

During the last several thousand years, the most significant expansions have started in Central Asia, thanks to technological developments in pastoral economies. Agriculture did not do well on the Asian steppes, but domestication of the horse afforded Eurasian pastoralists unprecedented advantages in migrations and military conquests. Many waves of migration started from the Kurgan region and had a profound impact on European and Asian history. The first expansion toward southern Asia probably occurred between 3000 and 2000 B.C., heading for Iran, Pakistan, and India via Turkmenistan. This passage appears to have contributed to the disappearance around 1500 B.C. of the Indus Valley civilization, which had produced the magnificent cities of Harappa and Mohenjo-Daro. At the same time as these nomadic expansions, there were dynasties related to the Indo-Europeans throughout the steppes as far as the Altai Mountains.

Around the third century B.C., groups speaking Turkish languages of the Altaic family, like the Huns, began developing new weapons and strategies. In the next centuries they threatened empires in China, Tibet, India, and Central Asia, before eventually arriving in Turkey. In 1453 Constantinople and the Byzantine Empire fell before their armies. The conquests of their descendants continued into recent times, with expansions to Europe and North Africa. Genetic traces of their movements can sometimes be found, but they are often diluted, since the numbers of conquerors were always much smaller than the populations they conquered. In Turkey and the Balkans, the furthest point permanently settled by these Mongolian nomads, no clear genetic trace of their origin has been found, but genetic investigations are limited. Further expansions of these Eurasian nomads recorded by history are those of the Avars, Scythians, and all of the barbarians who put an end to the Roman Empire. Most of the earlier conquests are poorly known.

Genetic analysis indicates a major expansion began from near the Sea of Japan—possibly even Japan itself—but it is difficult to date. It may have been very early. According to our archeological knowledge, it could have occurred 11,000 or 12,000 years ago, coinciding with or even preceding the date of ceramic development. We cannot entirely disallow that ceramic technology subsequently spread from there to the Middle East. Pottery was important for the preservation of food, and it would be necessary to know more about the dates of ancient ceramics in a wider area around Japan. The demographic growth determined from statistical analyses of archeological sites in Japan indicates that a demographic maximum was not reached until 4,000 years ago—another possible date for the expansion the genetic data indicate.

Certain mutations conferring malarial resistance, often in the heterozygous state, are concentrated along the Mediterranean coast and the Pacific Ocean where malaria has been a serious disease. In the tropics, and even in some temperate climates, malaria is probably the most serious human disease. Several genetic mutations confer a selective advantage in malarial areas, like the thalassemias and sickle cell anemia. It has even been possible to track ancient Greek, Phoenician, and Malayo-Polynesian migrations by studying DNA markers of the thalassemia genes, as well as other diseases conferring resistance to malaria.

The northern Andes experienced an important population expansion, which probably began in Mexico. One route toward the Brazilian plain might have been sustained by the cultivation of manioc, which was able to grow in tropical forests. We have already seen manioc's extraordinary success in Africa, where it replaced cereals that had previously sustained the Bantu expansion. The progress of agriculture was slow in northern Mexico, where deserts delayed its spread to North America until about 2,000 years ago.

In Australia, the great expanses of coastal forests and the internal deserts did not favor agriculture, which started prospering only after James Cook reached it at the end of the eighteenth century. In New Guinea, now separated from Australia by water, agriculture could thrive especially in the internal highlands, which witnessed agricultural developments early and over many millennia. More recently, the coasts were also colonized, by Malayo-Polynesians; and for most of human history New Guinea, an island smaller than Australia, had a larger population.

It is obvious that expansions have punctuated the last 100,000 years of modern human evolution and that their genetic trace can be seen in principal components maps. In general, expansions are determined by the invention and use of new technologies that stimulate demographic growth and eventual migration. Increased food production, for example, can spur demographic growth, which pushes new populations to migrate and to occupy and cultivate new territory. Innovations in transportation can also aid migration. Likewise, military power was advantageous, or even necessary, during several late episodes of expansion when populations were forced out of an area. But military action was rarely the principal cause for demographic expansion and was not a dramatic source of genetic migration. If military superiority helps a small force subjugate a larger population, the genetic effect is trivial, though the cultural effects are often profound. Today, however, we can see an amplification of sex differences, comparing mitochondrial DNA and the Y chromosome. The genetic consequences of population expansion depend on the ratio of the number of migrants to the number of inhabitants in the occupied region.

Let's take the case of primitive farmers moving into an area inhabited by hunter-gatherers. The latter would be living at a lower saturation density, and would reproduce very slowly (about one child every four years) resulting in near zero growth. Hunter-gatherers are seminomadic and must carry everything, including small children, when they move, and this has been recognized as a major cause for a low fertility rate, which is just sufficient to balance mortality. The Pygmies maintain a sexual taboo for three years after the birth of a child. Such taboos, less drastic, are also found in other African populations. Cultivation and animal breeding led to a population density a thousand times greater, and also ended the nomadic life that had limited the number of manageable children. In more sedentary societies, children in large numbers are an advantage as workers, and caretakers of the elderly. Agricultural societies can thus grow very rapidly. In general, farmers believe that they are superior to hunter-gatherers. Marriages between the two are often permitted, but the most common rule is that only farmer males are allowed—and only in certain societies—to take Pygmy wives, because they are believed to be more fertile, and are much less expensive to marry (wives are bought from their parents in most of sub-Saharan Africa). The reverse situation is not socially acceptable. A wife can move up, but not so easily down, in social status (a rule called hypergamy by anthropologists). If farmers are initially outnumbered when they enter a new area, they reproduce more rapidly than local hunter-gatherers, and quickly outgrow them. As they rapidly reach a higher saturation density, they have a genetic advantage over hunter-gatherers, since the ultimate genetic composition of a region depends on the relative numbers of the various genetic types.

Pastoral nomads figure somewhere in between sedentary farmers and hunter-gatherers in terms of population density. They often live in camps outside farmers' villages or towns, but can easily multiply and expand, having few reasons to control reproduction before settling down. They frequently build military force to protect their herds, and this often allows them to acquire control over large groups of farmers. Aryans, the nomadic pastoralists who occupied the Indian subcontinent, formed a society of many castes. These castes were organized in a rigid hierarchy, and were—still are, at least in rural India—strictly endogamous (marriage between castes was prohibited) or at most hypergamic (women were allowed to marry into a higher caste than that of their origin). Original Aryans formed the highest, or Brahman, caste, which provided the priests, philosophers, and practical leaders in all Hindu societies. Power and authority derived from social status, not numbers. Aryans spoke and spread Indo-European languages to Afghanistan, Iran, and India. Extension of the name Aryan to include Europeans and in particular Germans, supposed to be the original Indo-Europeans, is a fantasy that began in Germany and was especially dear to Nazi theorists. In Sanskrit, the old language of Indo-Iranians, *aryas* means noble, lord, ruler.

Every expansion produces different genetic gradients, as people spreading from the area of origin mix to different degrees with earlier settlers. We could not have seen the genetic influence of each separate migration without the help of principal components maps. In the near future, things may change. New developments of molecular genetics are making it possible to study more directly the migratory paths of single individuals throughout human evolution, allowing a more subtle dissection of expansions. But it will take time to accumulate the necessary body of data, especially at the current level of funding of research.

One might ask if the order of principal components can be given a meaning. It is likely that the first components correspond to the oldest events, since population sizes were smaller in the past, and initial genetic differences between populations were maximized by genetic drift. Each principal component measures the global genetic variation due to the genetic gradients that it detects. The more pronounced the gradient, the greater the total amount of variation explained by the component. In Europe, we can verify that there is a correlation between component rank and age. The fractions of variance explained by the first five components in Europe are 28 percent, 22 percent, 11 percent, 7 percent, and 5 percent. The first expansion can be dated to between 9,500 and 5,500 years before the present (BP). The second expansion is probably more recent, although we have very little archeological or linguistic information about the Uralic expansion. But if it is correct that the second PC is also influenced by the postglacial expansion of Mesolithics from the Basque region, then the average dating of the two components may be similar to that of the agricultural expansion. There is no major difference between the factions of variance corresponding to the first and second PC anyway. The origins of the Kurgan culture (the third PC) must be younger—perhaps 5,000 to 5,500 years ago at earliest. The Greek migrations suggested by the fourth component probably date from 2,500 to 4,000 years ago. It thus appears that the chronological order of the expansions is approximately reflected in the order of the principal components. As for the fifth component, the Basque culture, it reveals more of a population contraction caused by a long period of recent expansions external to the area, which the Basque culture has so far resisted, progressively losing ground. Experience with PC analysis shows that it is possible to observe an influence of the same phenomenon on different components. Here the second would be connected with an early expansion, the fifth with a later contraction due to other populations coming from the outside. But even if there is a correlation between the order of principal components and time, it is clear that PC analysis is not a method of dating.

Genetic Chronology

Principal components can be approximately compared to archeological strata, which formed the basis for a qualitative relative dating before the discovery of radiocarbon analysis. This later method allowed absolute dating, which depends on a physical measurement—the rate of ^{14}C's radioactive decay. The decay of the amount of ^{14}C relative to other types of carbon (^{12}C and ^{13}C, which are not radioactive and are therefore stable) is not influenced by temperature or other chemical or physical forces. It is therefore a physical clock, usable only on material containing sufficient carbon. I have mentioned that the method is flawed because one of its basic assumptions—that the amount of atmospheric ^{14}C remained constant in the past—is not completely true. Nevertheless, it has been possible to compensate for this factor, using dendrochronology (sequences of rings in ancient wood) for correcting ^{14}C datings.

Could we use a similar method for genetic dating? Until very recently, genetic dating has depended on employing a calibration curve. Geological and paleontological events that happened at known dates and could be held responsible for biological events recognizable in the fossil record (like the mammalian radiation and the disappearance of dinosaurs) were used to establish a "calibration curve." The age of the biological events, like the differentiation of mammals or the separation of the evolutionary line into chimpanzees and humans, was measured, for instance, by the number of differences in their proteins, or in nucleotides of specific DNA segments. Such differences, plotted against the dates of the corresponding events, form the calibration curve. In this way the date of the human-chimp separations was estimated at about five million years ago, and the separation of Africans from non-Africans gave a date of 143,000 years ago using mtDNA results. This date, however, the most recent estimate of the birth date of the so-called "African Eve," is not necessarily the date of settlement of other continents out of Africa, but rather that of the most recent common ancestor, which is likely to be earlier.

Attempts at introducing absolute dating methods in genetic chronology have used mutation rates as the clock. One difficulty is that mutation rates are usually very poorly known, and another is that all methods are based on a number of other assumptions, in particular patterns of growth rates, which are not well tested or testable. One recently introduced method lessens these difficulties by using genetic markers called microsatellites. They have an elevated mutation rate, which unlike all other mutation rates has been estimated rather accurately. A first estimate made with this method gave results very similar to those obtained for African Eve. We now have reasons to correct this date, because later observations have shown the mutation process of microsatellites is more complicated than originally assumed; taking into account the added complexities has nearly halved the date.

Previous estimates led to early dates for the expansion of modern humans—between 100,000 and 200,000 years ago—but did not take into account the unique dynamics of the expansion and the considerable population increase it caused. A number of recent independent genetic datings place the beginning of expansion from Africa close to 50,000 years ago, a date first suggested by anthropologist Richard Klein on the basis of archeological research. Klein emphasized the significance of a newer, more sophisticated Aurignacian stone tool that replaced the Mousterian one used by Neandertal and archaic *Homo sapiens,* including early anatomically modern humans living in Israel nearly 100,000 years ago.

Peter Underhill and Peter Oefner of Stanford are the main authors of a soon-to-be-published set of Y chromosome data, which will confirm these claims, buttress them with data from other genetic systems (Li Jin et al. 1999, Luis Quintana-Murci et al. 1999), and greatly enrich our understanding of the expansion. The first serious development in Africa most likely occurred in the east and south, and the first expansion probably went from East Africa to southern Asia and Southeast Asia. From there, expansion continued south to Oceania and north to China, Japan and Siberia, and eventually on to America. The coastal route must have been quite important. There was an expansion from East Africa to northeastern Africa, and then to central Africa and West Africa. The Red Sea and Suez were another much used passageway to Asia. Not surprisingly, Central Asia has considerable genetic variation: it was settled from many directions and contributed to numerous new expansions. Europe began to be settled 40,000 years ago, probably from many places of origin: from Morocco, Tunisia, and the Middle East and Turkey, through the Ukraine and even across the Ural Mountains.

CHAPTER 5

Genes and Languages

More than 5,000 languages are spoken today. A few are used by hundreds of millions of people, but the great majority have a very restricted distribution. Languages with only a hundred speakers or fewer are in danger of imminent extinction; many have disappeared already.

It doesn't take a linguist to know that some languages are more closely related than others. Spanish and Italian, my mother tongue, are obvious examples. I can get by in Spanish- or Portuguese-speaking countries without much difficulty. However, words that are identical or similar but have different meanings cause trouble. For example, *burro* means "butter" in Italian but "donkey" in Spanish; *equipaggio* means "crew" in Italian and *equipaje* means "luggage" in Spanish; *salire* means "to go up" in Italian, and *salir,* "to go out" in Spanish. We call these words "false friends," and fortunately there are not many of them. Italian, French, Spanish, Romanian, and so on, derive from a common source—Latin. Likewise, Germanic languages include Swedish, German, Dutch, Flemish, and English. The Slavic languages of eastern Europe are also quite similar. The resemblance between Sanskrit, a classical language in India, and some ancient European languages was well-known as early as the eighteenth century.

The study of Sanskrit provided the first linguistic clues to the relationships among what would become known as the Indo-European family of languages. Since then, many other linguistic families have been recognized (some linguists prefer to call these families phyla). Like plant and animal taxonomists, linguists have reconstructed trees illustrating linguistic relationships that they call "genetic"—equivalent to the word's use in biology. But linguists have had trouble reconstructing relationships above the family level—we have yet to agree on a single tree linking all the existing families. In fact, many linguists believe that the question of the unity or diversity of modern languages can never be answered. The difficulty rests with the rapidity of linguistic evolution.

Figure 11 shows the geographic distribution of language families recently proposed by Merritt Ruhlen. The least well-known languages, those of the Australian Aborigines and New Guineans, are more difficult to classify. But opinions about other families also differ, and bitter feuding has characterized much of historical linguistics over the past century. One of the most contentious issues among American linguists has been the classification of Native American languages.

At the beginning of the century, having noted the resemblance of many Amerindian languages, the linguist Edward Sapir and the anthropologist Karl Kroeber claimed that only a few Native American language families existed. Their hypothesis met considerable resistance from the majority of American linguists who strongly opposed such a unification. A new cycle of contention began after Joseph Greenberg of Stanford University published a book in 1987 called *Language in the Americas,* showing that the languages spoken by pre-Columbian Americans could be grouped into just three families: Eskimo-Aleut, Na-Dene (comprising languages that are spoken mostly in the Pacific Northwest, but also including Navajo and Apache), and Amerindian, which comprises most languages of North and South America. Greenberg's proposal agreed with the

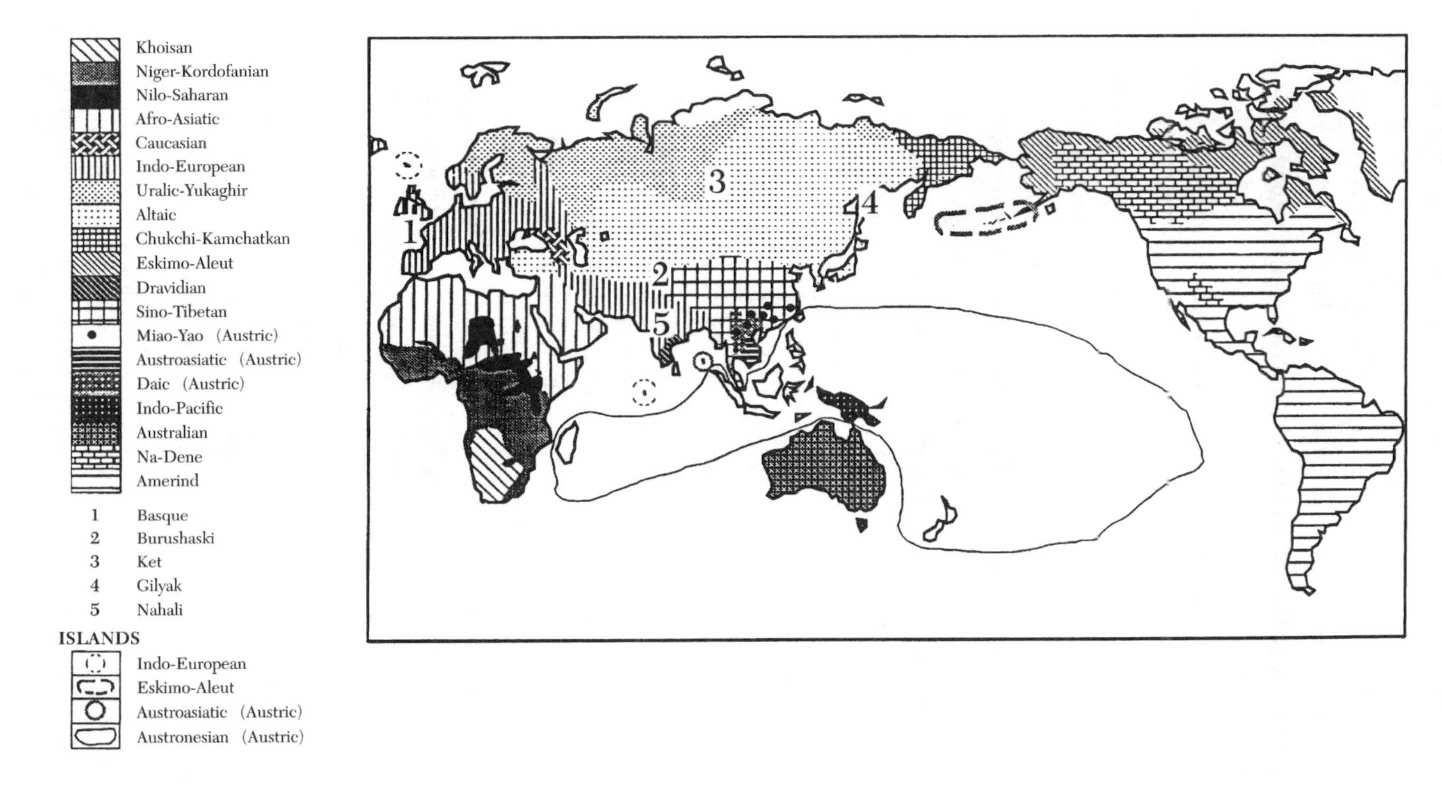

Figure 11. Geographic distribution of the seventeen linguistic families, and location of several linguistic isolates (drawn from maps in Ruhlen 1987, vol. 1).

classifications of American biologists Christie G. Turner and Stephen Zegura, who used measurements of modern and fossilized teeth, and blood groups and proteins, respectively. What's more, these three linguistic families seem to correspond with three major migrations suggested by archeological data. Amerindians appear to have come first, followed by Na-Dene speakers and finally Eskimos. The first group occupied all of America, while the second and third groups have remained near the Arctic, where they originated. We also found that Native Americans can be divided genetically into the three distinct groups Greenberg recognized on linguistic grounds. It must be said, however, that Amerindians are genetically extremely variable and that linguistic sub-groupings within the Amerindian family level do not correspond terribly well with the genetic results. Southern Na-Dene (Apaches and Navajos) are genetically similar to Northern Na-Dene, but the southern populations have absorbed genes from their Amerindian neighbors.

The Amerindians appear to have reached the Americas in a migration much older and more complex than those of later Na-Dene and Eskimo-Aleut speakers, and there may even have been more than one Amerindian migration. Genetic data indicate that Amerindians arrived at least 30,000 years ago, but this date may only represent an average of the most important migrations. In addition, it may be biased upwards if it is true that the first Amerindian migrations involved very few individuals, as some new data on Y chromosomes suggest. A strong founders' effect tends to increase the length of branches in neighbor-joining trees, discussed later, and thus also to exaggerate the time of first settlement calculated genetically.

The publication of *Language in the Americas* unleashed a new war between American linguists and anthropologists supporting Greenberg's thesis. A large group of linguists held a meeting and declared it impossible to recognize fewer than sixty or so taxonomic groups of Native American languages. Taxonomists can be divided into "lumpers" and "splitters": these synthetic and analytic tendencies probably reflect a fundamental dichotomy in the human spirit. However, in the case of Amerindian classification, methodological differences can explain much of the dispute, as Greenberg discusses in detail. I am not a linguist, but I find Greenberg's arguments most convincing. Furthermore, Greenberg has been through this before. Many years ago, he proposed a now widely accepted classification of African languages into only four families: Afroasiatic, including all Semitic languages and most languages in Ethiopia and North Africa; Nilo-Saharan, comprising languages spoken along the upper Nile and the southern Sahara; Niger-Kordofanian, which includes most central, southern, and West African languages, especially the Bantu ones; and Khoisan languages, spoken by Khoi-Khoi and San populations in southern Africa. Greenberg's classification endured a barrage of criticism when it was first proposed, and it is now widely accepted. A similar change of attitude toward the Amerindian classification will probably come with time.

An examination of some of the objections that Greenberg's colleagues have raised against his classifications can help us understand both the objective difficulties that afflict studies of linguistic evolution, and also the subjective ones, which are typified by Greenberg's attackers. Languages change very rapidly and it is terribly hard to establish clear connections between distant languages. Significant phonological and semantic changes to all languages occur over time. The magnitude of these changes complicates the reconstruction and evaluation of linguistic commonalities. Grammar also evolves, although usually sufficiently slowly to aid the recognition of more ancient linguistic connections. Under the pressure of phonetic and semantic change, a language rapidly becomes incomprehensible. Modern languages derived from Latin would not be understandable to a Roman of two thousand years ago—a thousand-year separation is often enough to render a language incomprehensible to its first speakers. After a separation of five to ten thousand years, the rate of recognizably similar words can drop to ten percent or less. Fortunately, certain words and certain parts of speech exhibit a slower rate of change and give us a better chance of discerning more distant linguistic relationships.

As for the problems caused by misguided methods, some anti-Greenberg linguists believe it is impossible to posit a quantitative relationship between any two languages. By disallowing reliable measurements, and by limiting the relationship between two languages only to "related or not related," the American linguists opposing Greenberg have ruled out the possibility of hierarchical classifications, an essential prerequisite to taxonomy.

Interestingly, this position completely contradicts the view of linguists who use sophisticated methods to measure linguistic similarity, based on the fraction of words from a standard list that have a detectable common origin. This approach was developed by an American linguist, Morris Swadesh. He suggested that the probability that a word will lose its original meaning stays constant over time. When the fraction of related words retained after a known period of time has been calculated (by observing the change from Latin to the modern Romance languages, for example), we can construct a "calibration" curve, allowing us to read the time elapsed since two living languages shared a common ancestor. This method, which has been called "glottochronology," uses a "linguistic clock" very similar to the "molecular clock" we discussed previously for genetics. In biology, we have the advantage of using many proteins or DNA sequences to obtain many independent estimates of the date of separation between two species. Unfortunately, there is no such variety and wealth of data in linguistics to strengthen our conclusions. Glottochronology is a less rigorous method than those used in biology, and is especially difficult to apply to distant comparisons when the fraction of cognate words more than ten thousand years ago is very small. The word lists cannot grow, because only a limited number of words change slowly; moreover, every word has its own rate of change, a fact neglected by glottochronology, which assumes a constant rate of change.

Other linguists insist that the resemblance between similar words in different languages be examined in light of "classical sound correspondences," which are very strict rules of phonological change. If these rules are not followed exactly, they say, two words cannot be considered "cognates," that is, one cannot determine if they share a common origin. Greenberg has answered them with an impressive list of exceptions to these rules within the Indo-European language family and others. He concluded that it would be impossible to establish the Indo-European language family if these rules were applied too rigorously. Fortunately, the Indo-European family was proposed and accepted before the theory of sound correspondences adopted its most rigid format.

Finally, some linguists believe that the parent language that gave rise to a family or, in general, a cluster of languages must be reconstructed in order to demonstrate a phylogenetic relationship among families. Here too, biology provides an analogy—when the "consensus" DNA sequence is calculated from the sequence of two modern species. The consensus sequence is the best guess of the ancestral sequence that would require the fewest changes to generate the diversity observed in a particular sample. But the search for consensus is less rigorous in linguistics because linguistic variation is much wider than biological variation, since only four nucleotides make up DNA. In biology, some proteins are so important to an organism that little or no change could be tolerated. Thus, many sequences of proteins change extremely slowly, and it is possible to prove their relationship without reconstructing ancestral sequences of millions or even billions of years ago. Knowledge of a proto-language may be helpful in comparative analyses, but imposing this exercise on all linguistic classifications is a serious limitation when so few proto-languages have been generated. Furthermore, reconstructions have a low probability of being entirely reliable. Greenberg's method avoids this impasse. It may be more subjective than is desirable, but it can go much further than other methods.

The classification of families by Merritt Ruhlen (a student of Greenberg's) appears to me to be satisfactory for comparing genetic and linguistic evolutions, as we will do in the next section. Defining a family does not appear to be an entirely objective task, but the distinctions between families, subfamilies, and superfamilies are mostly a matter of convenience and are unnecessary for certain purposes. What matters is the possibility of establishing a simple, logical, and hierarchical relationship. Unfortunately, most modern classifications stop at the level of families, of which there are as many as seventeen in Ruhlen's unifying system. There are some superfamilies, but, as already noted, modern linguistic methods have not yet generated a complete tree growing from a single source.

It is of interest to consider some of the proposed superfamilies, even if they are rather controversial. According to Ruhlen, Austric is a superfamily consolidating four families: Miao-Yao (spoken in pockets of southern China, and northern Vietnam, Laos, and Thailand); Austroasiatic, comprising Munda languages spoken in northern India and Mon-Khmer (spoken mostly throughout Southeast Asia); Daic (spoken widely in southern China and much of Southeast Asia), and Austronesian. There are some 1,000 Austronesian languages, spoken by about one hundred eighty million people, including Taiwanese aborigines and Malayo-Polynesians. The latter group ranges from Taiwan to Polynesia, parts of Melanesia, the Philippines, Indonesia, Malaysia, and as far west as Madagascar. The oldest Austronesian languages are spoken by Taiwanese aborigines. Whether or not this superfamily is accepted, it ties together a very wide geographic region, including Southeast Asia, both insular and non-insular areas, and a large number of islands in the two oceans separated by Southeast Asia.

Superfamilies extending over Europe and Asia are of particular interest. Two linguistic groupings in this region are current today and are closely related: Nostratic and Eurasiatic. Rejected at first by most linguists, they are slowly being recognized. The Nostratic superfamily, as originally described by various Russian scientists, includes the Indo-European, Uralic (spoken across the Uralic mountains), Altaic (widely spoken in Central Asia), Afroasiatic, which comprises many North African and also Semitic languages, Dravidian (currently spoken almost only in southern India), and South Caucasian families. One Russian linguist, Vitaly Shevoroshkin, showed that the Nostratic superfamily has strong similarities to the Amerindian group, as defined by Greenberg. The Eurasiatic superfamily proposed by Greenberg is similar to Nostratic, but it is different in the extension given to some families like Altaic and comprises smaller families like Eskimo and Chukchi, in addition to Japanese. Thus, Eurasiatic extends further east than Nostratic, but not as far southwest, as it does not include Afroasiatic and Dravidian, which, according to Greenberg, have an older origin.

One could continue building on this tree by adding an earlier branch, leading to the Nostratic/Amerindian group on one side, and to a new, older superfamily on the other, Dene-Caucasian. Sapir initiated this new grouping, but Sergei Starostin proposed it officially only a few years ago. The Dene-Caucasian superfamily includes essentially three families: North Caucasian, Na-Dene, and Sino-Tibetan. The last family of languages is spoken by almost a billion individuals (in China, India, Nepal, Burma, and also Southeast Asia, plus some isolates in Europe and West Asia), and is therefore the most populous one.

Thus, one could draw this approximate diagram:

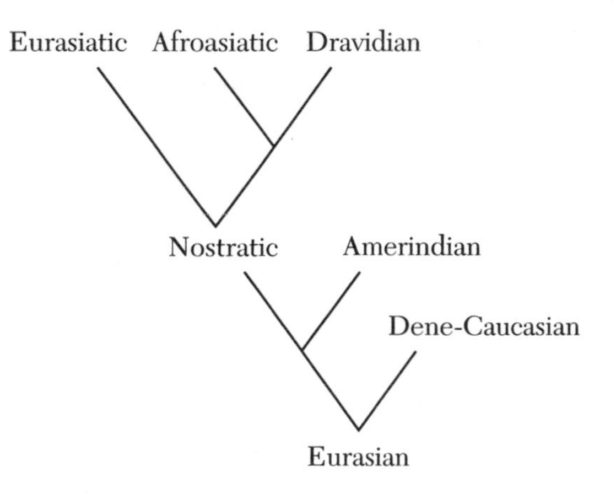

This hierarchy of superfamilies includes almost all of Europe, northern Africa, most of Asia, and all of America. It lacks only three African families, Khoisan, Niger-Kordofanian and Nilo-Saharan, as well as Australian (170 languages), and Indo-Pacific, a group of 700 languages, spoken mostly in New Guinea but also in neighboring islands and in the Andaman Islands near Malaysia.

There is, however, a small group of languages called *isolates,* which most linguists are unable to classify in any of the better-established families. The best known is Basque. Still spoken by approximately 12,000 French and perhaps a million and a half Spanish, this language is likely a relic of a pre-Neolithic period and is possibly related to the language spoken by Cro-Magnons, the first modern humans in Europe. But it has certainly changed enough that modern Basques and Cro-Magnons would not be able to communicate if they had the chance to meet. In fact they probably wouldn't even recognize their languages as related. Several linguists suggest a relationship between Basque and modern languages of the northern Caucasus. It is thus possible that one or more pre–Indo-European languages were spoken in Paleolithic Europe. Other linguists see even more encompassing resemblances among Basque, Caucasian, Sino-Tibetan, and Na-Dene languages. The last are spoken in the northwestern region of North America. Another claim is that Burushaski, an isolate spoken in a high valley in the Himalayas, is related to Basque and Caucasian. And, according to other linguists, Sumerian, Etruscan, and other linguistic "fossils" belong to the same ancient family, Dene-Caucasian. If the Nostratic/Amerindian group were added to the Dene-Caucasian group to form a hypothetical Eurasian superfamily, it may have extended across all Europe and Asia (except the southeast), and spread to the Americas. This giant superfamily later differentiated into several branches, and local twigs of this tree flourished and extended to regions far and wide.

These are interesting and hopeful hypotheses that need to be explored further. If we want to stay on absolutely firm ground, the situation is worse than simply lacking a reliable tree to link all modern languages: it is not even certain that all languages share a common origin. Most linguists consider both problems insoluble. It's a bit like trying to determine if all life on the planet had a single origin. (Many biologists believe in a single origin, since there is only one form of the twenty aminoacids found in proteins.) Greenberg noticed that there is at least one word all the linguistic families share: the root *tik*. It means finger, or the number one (a semantic shift, which requires no explanation). In other languages we find other semantic changes of this root, which also appear acceptable, "hand" and "arm," for example, or "point, indicate." In French, *doigt* (pronounced "dwa") and in Italian, *dito* (meaning "finger") come from the Latin root *digit*.

Extending this example, American linguist John D. Bengtson has proposed, with Ruhlen, about thirty other roots that are nearly as universal. But it will take a long time for other linguists to examine and accept these newest results. As might be expected, there are very few roots that are common to most languages. Most of them designate parts of the body, or are personal pronouns or small numbers (one, two, and three). It is not surprising that words conserved since the beginning of linguistic diversification are among the first words we learn: eyes, nose, mouth, and so on. But there are others that were certainly very important in Paleolithic life and have been preserved in many languages; "lice" is one example.

Comparison of Linguistic Families with the Genetic Tree

Even without a comprehensive linguistic tree, we can still compare our genetic tree to existing linguistic trees. There are some impressive similarities.

In figure 12, the language families have been drawn next to the populations that speak the respective languages. We see that a family corresponds to one or more branches in the genetic tree. Sometimes a language family is represented in the joint genetic-linguistic tree by only one branch, because the populations speaking these languages were grouped together in the genetic analysis. In effect they show great similarity, either genetic or ethnographic, and live near each other. An example is the Bantu subfamily, belonging to Niger-Kordofanian, which is genetically homogeneous and distinct from other African groups. Although the word Bantu designates a language group, it is also useful as a biological category. Other genetic groupings also have been corroborated by linguistic information; for example, southern Indians speak Dravidian languages and the Na-Dene speak Native American ones. Shared language families often point to a common genetic and ethnic background.

The genetic tree shown in figure 12 is composed of 38 populations, some of which are broadly grouped (e.g., Europeans, Melanesians). There are only 16 language families (we had no genetic data on Caucasian populations when we drew this tree). Therefore some

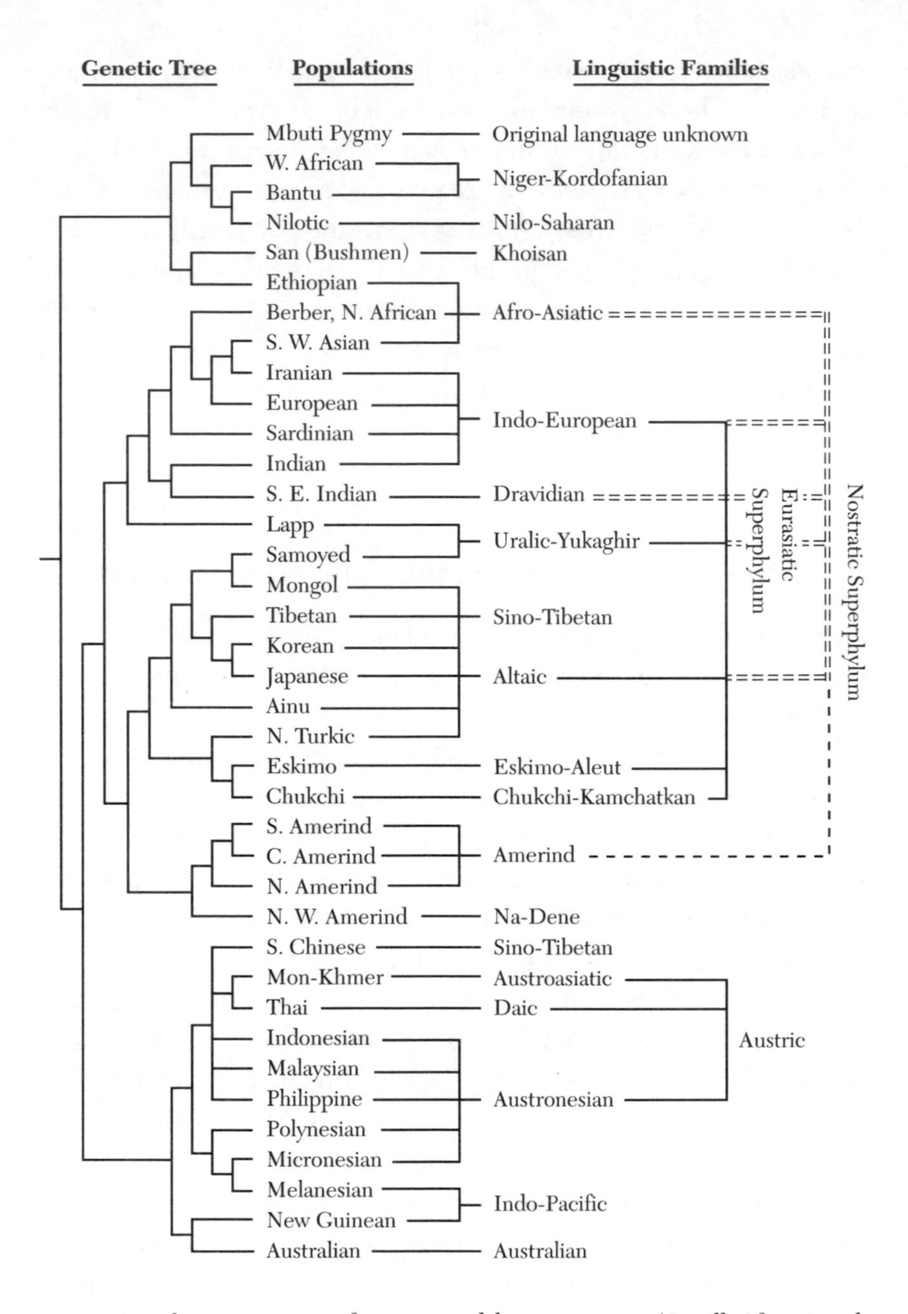

Figure 12. The comparison of genetic and linguistic trees (Cavalli-Sforza et al. 1988, pp. 6002–6).

populations in the genetic tree must and indeed usually do belong together to the same linguistic family. We can immediately note that populations that are adjacent in the genetic tree usually speak languages of the same family. Because of this we can use the genetic tree to help us date the approximate origin of a linguistic family. Most language families appear to have developed during a brief period, between 6,000 and 25,000 years ago.

There are, however, exceptions to the tendency of populations speaking related languages to be close in the genetic tree. Ethiopians, for example, are part of the African genetic branch, but they generally speak languages from the Afroasiatic family, which are widespread in North Africa and the Middle East, where the people are generally Caucasoid. Ethiopians are in effect a bit more African than Caucasoid genetically and more Caucasoid than most Africans linguistically. The Saami (Lapps) illustrate another exception to this rule: they are genetically Caucasoid but speak a Uralic language whose other representatives live mostly in northeastern European Russia and northwestern Siberia near the Urals. The Uralic people of Asia are generally Mongoloid, and the Lapps are a mix of Caucasoid (probably from Scandinavia) and Mongoloid (of Siberian origin), with a prevalence of the former. Even without looking at their genes, we can see this in the color of their skin and hair, and in the color and shape of their eyes, which vary in type from Mongoloid to Caucasoid.

There is a simple explanation for these disagreements between genetic and linguistic classifications. As we have explained, these two populations are the products of relatively recent genetic admixture: of Europeans and Siberians for the Lapps, and of Africans and Arabs for the Ethiopians. In the genetic tree, the populations are placed with the groups that contributed the greater proportion of genes. Extended admixture may also put them in a more isolated and somewhat intermediate position in the tree. The genetic effects of population admixture are much simpler and more predictable than linguistic change. Genes of a mixed population occur in proportions corresponding to those of its ancestral parental populations. But a genetically mixed population tends to preserve only one of the two original languages. Sometimes, the language of a mixed population will not change at all; more often, however, we find a few words or, sometimes, sounds borrowed from the other language. Grammar is more resistant to change than vocabulary. As for the origin of the mixtures that generated Ethiopians and Lapps, we know of intimate contacts between Arabs and Africans in Ethiopia during the first millennium B.C. An Arab-Ethiopian kingdom established a capital first in Arabia (in the Saba region) and later in northern Ethiopia (at Aksum). But earlier contacts may have occurred that took place too early for history to record. We also know that the Lapps have been in their present territory for at least 2,000 years. In both these cases, the shortage of written records limits our ability to know just how far back contact originated. In each case, the degree of genetic mixing established can vary and depends on the amount and mode of contact between two populations.

It is easy to calculate that the genome of a people who received a constant genetic flow of 5 percent per generation from its neighbors would keep only 70 percent of its original genome after three centuries. This is the average value of admixture for African Americans, who have retained 70 percent of their original genome and received 30 percent from mostly white settlers. If this flux continues with the same speed, African Americans will have little more than 10 percent of their original genes after 1,000 years of habitation in America. In the cases of Lapps and Ethiopians, their parent populations may have been in reciprocal contact for a long time (perhaps several thousand years), and degrees of admixture are greater than those observed in African Americans, who were in contact with whites for a much shorter period, and under conditions of strong social inferiority.

We can find still other interesting exceptions to exact correspondence of the genetic and linguistic trees. Genetically, Tibetans belong to the group of northern Mongols, but they speak a Sino-Tibetan language like the Chinese. The Chinese represented in our tree, however, originate from southern China and are genetically more similar to southern Mongols. History comes to our rescue in this case as well. According to Chinese historians, Tibetans are related to the northern Chinese. Starting from the third century B.C.

in northern China, pastoral nomads headed south to Tibet. Some remained nomadic shepherds, but most maintained their original language after migrating. The unification of China began around the third century B.C. under the short-lived Qin (pronounced Chin) dynasty, and was completed under the Han Dynasty, which reigned for the next four centuries. These two northern dynasties can be credited with the spread of their language beyond northern China into the entire country. During the following 2,000 years, this language naturally differentiated into several others. Nevertheless, a large number of ethnic minorities (55 are officially recognized, forming about 10 percent of the total Chinese population), especially in southern China, have preserved their original language and genes, showing their different origins. The large majority of Chinese (close to 90 percent) call themselves Han, but the original genetic difference between north and south is still clearly visible among them. Therefore, it is not surprising that Tibetans have conserved their northern Chinese genes, although they live in the south, while southern Han look genetically more like Southeast Asians. But they all speak Sino-Tibetan languages, of northern origin.

The Nostratic and Eurasiatic superfamilies are represented at the extreme right of the group of language families in the tree in figure 12. With few exceptions, they correspond with the deeper genetic branches that we have called North Eurasian (not labeled) uniting the Caucasoids, northern Mongols, and Native Americans. This branch begins near the second fission, where non-Africans split into Southeast Asians (including Australians and New Guineans), and North Eurasians. The most important exceptions to this correspondence are the two Sino-Tibetan speaking populations, Tibetans and Chinese, whose genetic and linguistic associations noted in the figure disagree. They both speak Sino-Tibetan languages but Tibetans associate genetically with North Eurasians, and Chinese with South Asians. As we have just explained, however, the disagreement is superficial, because the Chinese shown in our tree are from the south, and have only adopted their current language in the last 2,000 years, while they are genetically more closely related to the speakers of Austric languages, that is, people living in Southeast Asia with whom they are grouped in the genetic tree.

Figure 12 shows one other apparent discrepancy between the genetic and linguistic trees: Melanesians (from the Pacific islands closer to New Guinea) are genetically close to Southeast Asians and are linguistically assigned to the Indo-Pacific family. This is not entirely accurate, because the languages spoken in Melanesia (as well as in coastal parts of New Guinea) are in part Austronesian, and part non-Austronesian. The latter are mostly Indo-Pacific, a very heterogeneous family. In practice, Melanesia's situation is not a disagreement but an exceptionally complicated situation generated by several superimposed migrations that can be considerably clarified by a more detailed analysis.

The genetic tree in figure 12 still has some flaws, which are discussed below, but, because of the complexity of the relationship and the frequency of mixed origins of populations, a much more detailed representation is needed. The tree we published in 1988 shows that after the first split (separating Africans from non-Africans), the second branch separates Eurasians and Americans from Oceanians (Australian Aborigines and New Guineans) and Southeast Asians. By 1994, new genetic information showed that the second branching split Oceanians from all the rest; Southeast Asians actually split from Eurasians in the third branch. Admixture between Southeast Asia and the rest of Asia is probably the reason for this uncertainty, which has yet to be resolved because of the lack of adequate data from Southeast Asia.

Other difficulties have also intervened, because a popular tree reconstruction method developed by Naruya Saitou and Masatoshi Nei (1987), called neighbor joining, produces a genetic tree somewhat different from that of figure 12. In neighbor-joining trees, Europeans are attached to a short branch near the center of the tree between Africans and Asians. The most likely explanation is that Europeans received some of their genes from Asia and some (a smaller portion) from Africa. Using the neighbor-joining method, mixing between populations shortens the mixed branch and moves it toward the tree's origin. I believe that the conflicts between these different tree-building methods are related mostly to admixture between populations.

Mixing between North Asians and Africans contributed to the European makeup. Several European genetic traits are intermediate between the two parent populations. Africa probably contributed genes to Europe via a number of different routes, including the Middle East. This region is reachable directly from both Africa and Asia without crossing the sea, and may have been a starting point for the occupation of Europe by modern humans 40,000 years ago. Neolithic humans came from the Middle East 10,000 years ago. But we cannot exclude the possibility that modern humans also entered Europe from northwestern Asia. This may explain the relationship which several linguists claim exists between Basque and the Dene-Caucasian languages spoken in extreme western Europe. Survival of the evidence of old relationships may be due to the refuge offered by geography, especially mountains, to relic languages and populations.

A genetic tree built using a reasonably large number of microsatellite markers showed that Basques were more similar to the Hunza (who speak the Burushaski language) than to four other Pakistani populations investigated. This observation, a result of the magnificent research effort done in Pakistan by Dr. Qasim Mehdi and his colleagues, will need confirmation with a greater number of individuals and markers. In the meantime, it provides the first genetic indication of a relationship between these two peoples, for which some remote linguistic resemblance was independently postulated.

Why Is There a Similarity between Biological and Linguistic Evolution?

There are important similarities between the evolution of genes and languages. In either case, a change, which first appears in a single individual, can subsequently spread throughout the entire population. For genes, these changes are called mutations; they are passed from one generation to the next and can—over many generations—

increase in frequency and even eventually completely replace the ancestral type. The genome is conserved and well protected from outside influences. Genetic mutations are rare, and transmission from one individual to another occurs only from parent to child, while linguistic changes are much more frequent and can pass also between unrelated individuals. As a result, languages change more quickly than genes. In effect, if a word can resist change for 1,000 years, a gene can remain substantially unchanged for millions, and even billions of years. Despite these differences, there are two reasons for anticipating important similarities in the evolution of the two systems.

Let me start by emphasizing that there is no reason to think that genes influence the ability to speak one language over another. If there are any such differences, they must be small indeed. Modern humans possess the capacity to learn any language, and the first language learned is a function of the time and place of birth. All modern languages share a similar level of structural complexity—ethnic groups that live at a primitive economic level do not speak a more primitive language than wealthier groups. If there is any interaction between genes and languages, it is often languages that influence genes, since linguistic differences between populations lessen the chance of genetic exchange between them.

Linguistic evolution is a special type of cultural evolution, as we shall discuss more generally in the following chapter. How is it possible for these two very different systems to follow parallel evolutionary trajectories, or to "coevolve"? The explanation is quite simple. Two isolated populations differentiate both genetically and linguistically. Isolation, which could result from geographic, ecological, or social barriers, reduces the likelihood of marriages between populations, and as a result, reciprocally isolated populations will evolve independently and gradually become different. Genetic differentiation of reciprocally isolated populations occurs slowly but regularly over time. We can expect the same thing to happen with languages: isolation diminishes cultural exchange, and the two languages will drift apart. Even if glottochronological estimates of the time of separation are not always as exact as we would like, in general languages do diverge increasingly with time. In principle, therefore, the linguistic tree and the genetic tree of human populations should agree, since they reflect the same history of populations splitting and evolving independently.

Nevertheless, there are several major sources of divergence between genetic and linguistic trees. One language can be replaced by another in a relatively short time. In Europe, for example, Hungarian is spoken in the geographic center of many Indo European branches: Slavic, Germanic, and Romance; but it belongs to the Finno-Ugric branch of Uralic. The other languages of the same family are spoken in the northeast of Europe and in the west of Siberia. At the end of the ninth century A.D., the nomadic Magyars left their land in Russia, crossed the Carpathians and invaded Hungary, which had already been occupied by the Avars. The conquest resulted in a Magyar monarchy, which imposed its language on the local Romance-speaking population. The number of conquerors was large but did not constitute the majority of the population—perhaps less than 30 percent of the total. The genetic effect of this conquest was therefore modest, and further diluted by subsequent exchanges with neighboring countries. Today, barely 10 percent of the genes in Hungary can be attributed to Uralic conquerors.

Elsewhere, the Barbarian conquests following the collapse of Rome faced greater difficulties in replacing or modifying the language of the conquered, who were always more numerous than their invaders. The earlier inhabitants also usually possessed a higher level of socioeconomic organization and were able to retain their cultural identity. The Lombards probably originated in Sweden or northern Germany, but they began to conquer Italy in the middle of the sixth century A.D. About 35,000 warriors, coming from Austria or Hungary, rapidly occupied most of Italy, except for the extreme south, and established a powerful state that lasted until the eighth century, but had no significant effect on the local language. The same is true of the Franks, a German population that played an important role in French political history without affecting their language. But in England after the fall of the Roman Empire, the Anglo-Saxons—Roman mercenaries of Germanic origin—did succeed in imposing their language, after establishing political control around the sixth century. The British Isles have witnessed dramatic linguistic change in a very short time. The indigenous population spoke pre–Indo-European languages unknown to us today. In the last millennium B.C., Celtic languages were spread throughout most of Europe, from a center probably located between Austria and Switzerland. At the time of Rome's conquest, Celtic languages were spoken throughout most of the British Isles. The Romans imposed Latin, which was followed by the adoption of Anglo-Saxon. Norman invaders ultimately brought many French words into English after 1066.

Another important replacement occurred in Turkey at the end of the eleventh century, when Turks began attacking the Byzantine Empire. They finally conquered Constantinople (modern Istanbul) in 1453. The replacement of Greek with Turkish was especially significant because this language belongs to a different family—Altaic. Again the genetic effects of invasion were modest in Turkey. Their armies had few soldiers and even if they sometimes traveled with their families, the invading populations would be small relative to the subject populations that had a long civilization and history of economic development. After many generations of protection by the Roman Empire, however, the old settlers had become complacent and lost their ability to resist the dangerous invaders.

In general, the survival of a language like Basque or Burushaski is more likely to happen in *refugia* (isolated places—like mountainous regions—resistant to invaders). A strong social identity also helps to retain a population's language.

Examples of language replacement are not limited to Europe, but because Europe's written history is quite long, the most recent replacements are documented to a unique extent. The Aryan invasions of Iran, Pakistan, and India brought Indo-European languages to Dravidian-speaking areas. The great geographic discoveries of the Malayo-Polynesians, extraordinarily skillful navigators, led to the diffusion of their Austronesian languages to parts of New Guinea, Melanesia and Micronesia, and Polynesia. To the west, Austronesian languages spread as far as Madagascar near the African coast. The Polynesian migrations had lesser genetic consequences in Melanesia and New Guinea, which were already occupied. The genetic-linguistic mosaic of Melanesia is very complex, reflecting a 5,000-year history of migration and admixture of different people. But when the latest Austronesian migrants—passing Melanesia and central Polynesia—reached eastern Polynesia, starting about 3,000 years ago, they still appeared nearly Mongoloid, because they had not had time to mix with the Melanesians.

Exploration enthusiasts will be happy to learn that, from a genetic viewpoint, it is still impossible to exclude the possibility that South Americans contributed in some ways to eastern Polynesia, as Thor Heyerdahl implied in his voyages with the *Kon-Tiki.* The genetic difference between Mongoloids and Amerindians is not sufficient to say exactly if and how South Americans may have contributed to Polynesia. Recently discovered genetic Amerindian markers could undoubtedly provide a clearer answer to these questions.

The total substitution of one language for another occurs more easily under the pressure of a strong political organization of newcomers, as witnessed in the Americas. Otherwise, the separate languages spoken in nearby countries can remain relatively unaffected for thousands of years, even when the genes of neighbors experience a partial, and sometimes major, substitution. It is difficult to quantify the extent of substitution that has occurred in the Basque genome through admixture with neighboring populations, but it must be considerable. However, given the length of time during which Basques were exposed to gene flow from neighbors, especially farmers who arrived in the area some 5,000 years ago, the gene flow per unit of time was small. There were perhaps only one or two mixed marriages per thousand each generation. By contrast, near complete genetic substitution without language replacement probably did occur in the Hadza and Sandawe. These two populations from Tanzania speak Khoisan languages, but their genes are unlike South African Khoisan. Both groups are quite small and must have lived among Bantus for quite a long time. Bantus probably arrived in the general area about 2,000 years ago. A population isolated among other, different ones, undergoing a gene flow of 5 percent per generation over 1,000 years, could result in the replacement of 87 percent of the population's original genes, and 98 percent over 2,000 years. The Hadza and Sandawe were hunter-gatherers and were thus separated from Bantu farmers by socioeconomic factors sufficient to preserve their own languages but insufficient to prevent genetic exchange with their neighbors. It must be admitted, however, that it is difficult to exclude the opposite hypothesis: that both the Hadza and Sandawe have basically maintained their original non-Khoisan genetic background, and have changed their language to a Khoisan one because of original contact with Khoisan speakers. Eventually, however, this contact disappeared after Khoisans retreated south. There are many such examples of language replacement with little gene replacement in the very recent expansions of Europeans that followed the introduction of transoceanic travel, resulting in the adoption of the invaders' language. The reverse also happened: Finns speak a Uralic language but they have very little, perhaps 10 percent, of Uralic genes (still to be confirmed with more powerful markers than those on which this estimate is based). It is possible they originally spoke a language of the Balto-Slavic subfamily, when they settled in Finland, a very vast country that must have been inhabited by a low density of Uralic-speaking hunter-gatherers or nomadic pastoralists, probably kin of the Saami still living in the northern part of Finland. As discussed earlier, there is genetic evidence that the original farmers who settled in Finland perhaps 2,000 years ago were a very small population, perhaps 1,000 or so. This is inferred from strong evidence of genetic drift, especially for certain genetic diseases. The new settlers probably joined a good number of native inhabitants, and peaceful contact with them helped the immigrants settle and spread. The process was facilitated by learning the natives' language, and eventually adopting it. Most likely there was little genetic exchange between the two.

In summary, the replacement of languages is not the only force that disturbs the parallelism between genetic and linguistic evolution. Genetic change due to gene flow from neighbors into a small group can be another one. Deeper analysis and, especially, historical information can frequently help to distinguish between the various explanations. It is remarkable that, despite the opportunity for genetic and linguistic replacement, we can still find sufficient coherence in the modern linguistic and genetic jumble to reconstruct a common tree for the two evolutionary tracks. But the rate of disappearance of traditional languages, a serious loss which is difficult to fight, is so high that this investigation may become impossible in a few generations.

Interpreting the Great Human Expansions on the Basis of Genetic and Linguistic Data, Particularly in Asia

We have already observed that, judging from the genetic tree, most linguistic families appear to date anywhere from 6,000 to 25,000 years ago. Some families are older; based on their time of colonization, the Indo-Pacific languages of New Guinea and those of Australian Aborigines may be older than 40,000 years. In this case the definition of families is aided by the great geographic isolation of these two regions, an island and a continent.

Khoisan languages must also be very old, given the uniqueness of some of their characteristics (e.g., the presence of click sounds), but it is difficult to know precisely how old. I would not be surprised if the ancestors of these peoples were responsible for the first expansion from Africa to Asia. There is some support for this perhaps surprising, certainly titillating, hypothesis, other than the possible antiquity of Khoisan languages. According to some anthropologists, the Khoisan people who now live in southern Africa once lived further north, in East Africa or perhaps even northeastern Africa. If this was the case between 50,000 and 80,000 years ago, then they were in the best position to expand to Asia. It is true that of all Africans, as we saw in figure 3 in chapter 3, East Africans are most similar to Asians. This is encouraging, but it might also be due to more recent migrations between East Africa and Arabia, which certainly happened much later in both directions. It is also worth noting that Khoisans bear some physical resemblance to East Asians, notably in their long-drawn eyes and large round heads. They also show a remarkable genetic similarity to western Asians (but not to

East Asians, in spite of their superficial resemblance for facial traits). At the time of writing, the first results of a major breakthrough in the study of Y chromosomes promise to give us some much anticipated answers to these problems.

Let us consider, in parallel, both genetic and linguistic data in other parts of the world, as they might illuminate the oldest expansions. There clearly must have been major expansions and migrations in Asia, soon after the arrival of modern humans. The principal components of Asia clearly indicate some regions that may be centers of major demographic developments. The first five PCs give us, in order: (1) a region of northwest Iran, south of the Caspian Sea, bordering Iraq on the western side and Turkmenistan on the northeastern side; (2) Southeast Asia; (3) the region around the Sea of Japan, including Japan, Korea, Manchuria, and northeastern China; (4) northern India; (5) Central Asia. One also notices a major genetic gradient of the first PC extending between east and west. It must have been generated by several migrations in eastern and western directions, of which there are many indications in history and in immediate prehistory. Obviously, there must have been many similar ones before. Also the Y chromosome analysis indicates that there must have been several different demographic developments in Asia, leading to the settlement of the three continents: Oceania, Europe, and America, in this order. It gives some idea of main migratory paths, but the data are not yet to the point of giving the precision one would like to reach. Recent research by Li Jin et al. (1999) comparing the Y chromosome and also a narrow but highly variable region of chromosome 21 shows that there must have been more than one major migration from Africa to Asia. The Y chromosome is particularly informative on older migrations, as shown by recent work of Peter Underhill and Peter Oefner, who use a new technique of discovering genetic variation called DHPLC. At the time of writing they have found 165 genetic variants of the Y chromosome, which can be divided into ten major groups (called haplogroups, and numbered from I to X). The first three are the oldest and originated in Africa, but the third born, III, migrated from

Africa to Asia. The other seven, from IV to X, originated in Asia and several of them are found also in Oceania, Europe, or America. They also correspond to centers of expansion indicated by the PCs: the VI and IX haplogroups correspond to the south Caspian. These two haplogroups may well include the whole Middle East, and may be a composite of several centers of expansion at different times, including earlier expansions from and to North Africa, and the development of agriculture. The VII and VIII correspond to the Southeast Asian center; the IV to the expansion from the Sea of Japan; the V to northern India. The expansion from Central Asia (the fifth PC) probably corresponds to a later branch of the V haplogroup. One can also assign, at least tentatively at this stage, a linguistic family or superfamily to each expansion center: Greenberg's Eurasiatic to the south Caspian region, Austric to the second PC center in Southeast Asia, Dene-Caucasian to the eastern center of the third PC.

The center of origin of Dravidian languages is likely to be somewhere in the western half of India. It could be also in the south Caspian (the first PC center), or in the northern Indian center indicated by the fourth PC. This language family is found in northern India only in scattered pockets, and in one population (Brahui) in southern Pakistan. It was spoken earlier further west—certainly in Elam (southwestern Iran), and possibly in the Indus valley (eastern Pakistan). The major group of residual Dravidian languages is spoken, as is well-known, in the south of India. It may seem strange to place the origin of a language family in an area in which almost no language of the family is represented today. But it is a reasonable assumption that this family was removed from its place of origin by the arrival in Pakistan and northern India of Indo-European speakers, 3,500 to 4,000 years ago. The effectiveness and cruelty of the Indo-Europeans' war against earlier settlers of India is told in vivid images in battles described in the *Mahabharata*.

The fifth PC indicates a center of expansion further north, approximately in the Altai region, and a natural suggestion is that this is the center of origin of Altaic languages. Two major expansions appear to have originated from this area, both late: by Mongols (third century B.C.) and by speakers of Turkic languages (beginning with the eleventh century A.D.). There surely could have been many other expansions before. In a period preceding the Mongolian and Turkic expansions, the three millennia before the Christian era, the area was in part controlled by Indo-European speakers (Tocharians).

Interestingly, the first suggestion of a reverse migration from East Asia was a hypothesis by Michel Hammer on the basis of a peculiar Y chromosome mutant he discovered in Japan and also in Africa. We found other mutants supportive of Hammer's observation. The origin of the specific Hammer mutant was not necessarily in Japan, but the migratory chain seems reasonable and independent from others. The haplotypes involved may be the Asian branch of the III Y-chromosome haplogroup, and the IV haplogroup. It is possible that this expansion may have brought the oldest Eurasian linguistic family (called Dene-Caucasian by Starostin) across Asia to reach Europe, perhaps 40,000 years ago. This family is represented in *refugia* by linguistic isolates like Basque, a Caucasian family, Burushaski in northern Pakistan, Ket (on the Yenisei River in central Siberia, possibly ancestral to the Na-Dene family in northwestern America) and some other extinct, less clearly related languages (like Sumerian and Etruscan). Two major families also belonging to the Dene-Caucasian superfamily, Sino-Tibetan and Na-Dene, survived over more extensive regions. They were nearer to the postulated region of origin, and must have survived in larger numbers than the other branches of the family.

Dene-Caucasian may have been the Eurasian superfamily par excellence until perhaps 20,000 years ago, which some linguists have given as the time of origin of Nostratic. Nostratic may have been a late branch of Dene-Caucasian. It started growing perhaps 10,000 to 20,000 years ago and gave rise to the families which largely replaced Dene-Caucasian in Eurasia: Indo-European, Uralic, Altaic. Greenberg's Eurasiatic family lengthens the northern Asian range of Nostratic to the east. It excludes Afroasiatic, whose geographic origin is obscure, but is earlier than Eurasiatic and originated in Africa, according to Greenberg.

Indo-European remains the most studied language family. Attempts to determine its place of origin have given incredibly variable results. Many locations have been proposed, ranging from Germany as far as the northeastern Caucasus and from the Baltic States to Suez. Some hypotheses are even wilder. Not long ago, one of the most popular theories was proposed by the archeologist Marjia Gimbutas, who postulated an origin above the Black Sea and associated the earliest speakers of Indo-European with the Kurgan culture of the Asian steppes. But when Gimbutas published her hypothesis, the Kurgan dates were poorly known. She assumed 3,000 to 3,500 years B.C., a date which was rejected as too old by English archeologists. Gimbutas's dates appear to have been vindicated by new excavations, which have also shown that horses were probably domesticated and mounted at that time and that war chariots were built in this area.

In 1987, Colin Renfrew proposed that Indo-European languages were conducted north by the Neolithic farmers of the Middle East. In chapter 4, I mentioned his influential book, which corroborated our hypothesis that Neolithic agriculture spread by a demic and not a purely cultural process. It is tempting to champion the correspondence between the spread of Indo-European languages and the diffusion of agriculture, which geography brings clearly to light. However, in my discussions with Albert Ammerman, my archeological collaborator in the initial research on the spread of farming and farmers, we avoided linguistic correlations because archeology cannot tell us about them in the absence of a written record. Nevertheless, on the basis of theoretical anthropological considerations, archeologist Renfrew came courageously to the conclusion that Indo-European was spread by Middle Eastern farmers.

I learned about Renfrew's hypothesis before he published it, on the occasion of a visit to Cambridge. Further connection between the spread of agriculture and language came to mind when I learned from linguistic literature that the language written in a cuneiform script around 5,000 years ago in the region of Elam (southwestern Iran) was Dravidian. Both Renfrew and I independently suggested that Dravidian may have originated in the Middle East and been spread by mideastern farmers east toward Pakistan and India. But in the last section of this book I tried to shift the origin of Dravidian away from the Fertile Crescent further east, either to the south Caspian, eastern Iran, or northern India. It seems very reasonable to assume that agricultural developments helped spread the languages spoken by the first farmers. This must have happened repeatedly, and we will see other examples. But agriculture developed no earlier than 10,000 years ago, and therefore the relevant linguistic families are late ones. If Greenberg is right in stating that Dravidian as well as Afroasiatic is older than Eurasiatic, then the center of origin of Dravidian is not necessarily connected with the Middle East, and may be further to the east.

Another interesting question related to the difficult problem of centers of origin of linguistic families arises with Renfrew's hypothesis that the Indo-European languages originated in Turkey, and then spread into Europe with Neolithic peasants. Obviously, all immigrants bring their language with them, and have no reason to learn a new one if they fail to encounter anyone in their new territory. It is worth pointing out that the inhabitants of Europe before the agricultural expansion (often called Mesolithics) usually had a very low population density. Since they were hunter-gatherers, they may have preferred living in areas that were geologically different from agriculturally suitable land. These earlier inhabitants and the new settlers, the Neolithic farmers, did not therefore have much contact, especially at the beginning of the agricultural expansion, when the density of farmers was lower than at later stages.

Renfrew's hypothesis, if correct, provides a date for the dissemination of Indo-European languages equivalent to that of the initial spread of farming, around 9,500 to 10,000 years ago. This date may seem to be problematic, since old linguistic estimates (although very approximate) suggested an earlier date (6,000 years ago). Moreover, this latter date would fit more comfortably Gimbutas's hypothesis of a Kurgan origin 5,000 to 5,500 years ago (kurgans are tombs in the form of mounds, which have been quarries of art objects in southern Russia). As we shall see, however, there is no necessary contradiction between Gimbutas and Renfrew. On the contrary, Alberto Piazza and I believe that their proposals reinforce each other. If we accept this idea, it may be useful to refer to the original Indo-European spoken in Turkey 10,000 years ago as the primary Indo-European, pre–proto-Indo-European, and to that spoken 4,000 to 5,000 years later in the Kurgan region as secondary or proto-Indo-European.

It is clear that, genetically speaking, peoples of the Kurgan steppe descended at least in part from people of the Middle Eastern Neolithic who immigrated there from Turkey. To arrive north of the Black Sea farmers from Turkey may have expanded west of the Black Sea, through Romania, and/or along the eastern coast of the Black Sea. Shortly after their arrival, these Neolithic farmers domesticated the horse, which was not as abundant elsewhere, and developed a predominantly pastoral economy. This allowed them to survive and even prosper in an environment ill suited to an exclusively agrarian life. This adaptation took time, but with the first development of bronze (around 5,000 years ago), they were on the brink of an expansion. They had food, a means of transport, and powerful new weapons. Actually, the Kurgan region extended fairly widely, and generated many expansions after this first one, over the next 3,000 or 4,000 years. The very first area of origin may have been between the rivers Volga and Don, but there were many expansions, both eastward to Central Asia and westward toward Europe. Kurgans are found over much of the steppe in both western and eastern directions.

The eastward expansion may have been first. It led east and south through Central Asia toward Iran, Afghanistan, Pakistan, and India, generating the "Indo-Iranian branch" of Indo-European. These languages later completely replaced almost all the Dravidian languages previously spoken from Iran to Pakistan and in northern India, but not those in southern India. Most inhabitants of India are Caucasoid, even if their skin is darker than that of northern Europeans. Populations in the south that speak Dravidian languages are genetically slightly different from, and darker than, northern Indians. At least three ethnic layers are superimposed in this part of the world. The oldest and the most limited (the pre-Dravidians, or Australoid) have unfortunately not been studied in detail. They are said to resemble Australian Aborigines in some respects; the similarity can only be superficial, but these people are likely to be more or less direct descendants from first African immigrants. As for the Dravidians, they were most probably the first Neolithic farmers of India, but it is unclear where they actually came from—perhaps from the Middle East, as originally hypothesized both by Renfrew and myself, or perhaps from northern Iran or northern India, as outlined above. Unfortunately, not much is known about the development of Indian agriculture.

The expansions in the opposite direction, westward, toward central and northern Europe, generated, one after the other, the Celtic, Italic, and Germanic branches of Indo-European languages. Northward expansions may have originated the Balto-Slavic expansion, which was perhaps the last. The southward expansion was less successful, as the area was already heavily inhabited, but from the second millennium B.C. there were various Indo-European speaking peoples and dynasties in Turkey and the Middle East—the Hittites and the Mitanni—whose probable origin was from the Kurgan.

That Gimbutas's and Renfrew's ideas seem more reasonable in combination than either is alone was confirmed by a recent study of the Indo-European language tree. We conducted this research using material published in 1992 by two linguists, Isidore Dyen and Paul D. Black, and the statistician Joseph B. Kruskal, to make the first complete, quantitative analysis of the similarities between Indo-European languages. The data published are the frequency of common origin of two hundred words in some six dozen Indo-European languages. All possible pairs of languages were compared and the similarity of each pair evaluated by calculating the percentage of words that showed common origin by classical linguistic criteria. For instance, "mother" in English and *mère* in French have common origin, while "head" and *tête* do not. The words were those of a standard list used for "glottochronology," a method for dating the separation of languages. Their statistical method is multidimensional scaling, a sophisticated type of principal components analysis. Using their data we applied two modern methods of tree reconstruction developed for genetic studies, and obtained easily reproducible trees that, interestingly, correspond reasonably well to August Schleicher's original one. The biggest difference is the position of the root, which as always is the most difficult to assess.

The most important groups of Indo-European languages are the following subfamilies: Germanic (which includes English and the Scandinavian languages), Italic (issuing Latin, among other languages spoken in Italy, in the first millennium B.C.), Balto-Slavic, Celtic, Greek, Indian, and Iranian. Most linguists consider Indo-Iranian a single branch, although Dyen and his colleagues say they are distinct. In our tree, several languages have an early, separate origin: Albanian, Armenian, and, later, though somewhat less clearly, Greek. Extinct languages like Hittite and Tocharian could not be considered in our analysis. The same tree was obtained by two other major methods of reconstruction and is shown in figure 13.

It is reasonable to think that isolated languages like Albanian and Armenian (and with less evidence Greek) originated with the first wave of Neolithic farmers from Turkey. Their greater age with respect to other families is responsible for their early position in the tree. They are also geographically closest to Turkey.

Our analysis groups the Indian and Iranian languages into one Indo-Iranian group, in the classic tradition of Indo-European studies, but slightly contradicting the results of Kruskal and his colleagues.

The later branches are likely to be derived from the second wave of Indo-European migrations from the Kurgan area, central European ones from its western part, and the Indo-Iranian branch from the eastern part. The order of branching in the tree is interesting: that of Celtic, Balto-Slavic, Italic, and Germanic subfamilies corresponds reasonably well to their geographic distance from the center of origin. Here the first branch of the tree is the Celtic, which is still spoken in the extreme western areas of Europe, and therefore is the most distant one from the Kurgan area. The next split generated the Italic-Germanic and the Balto-Slavic branch. The Italic branch settled in the southwest of Europe; it was unable

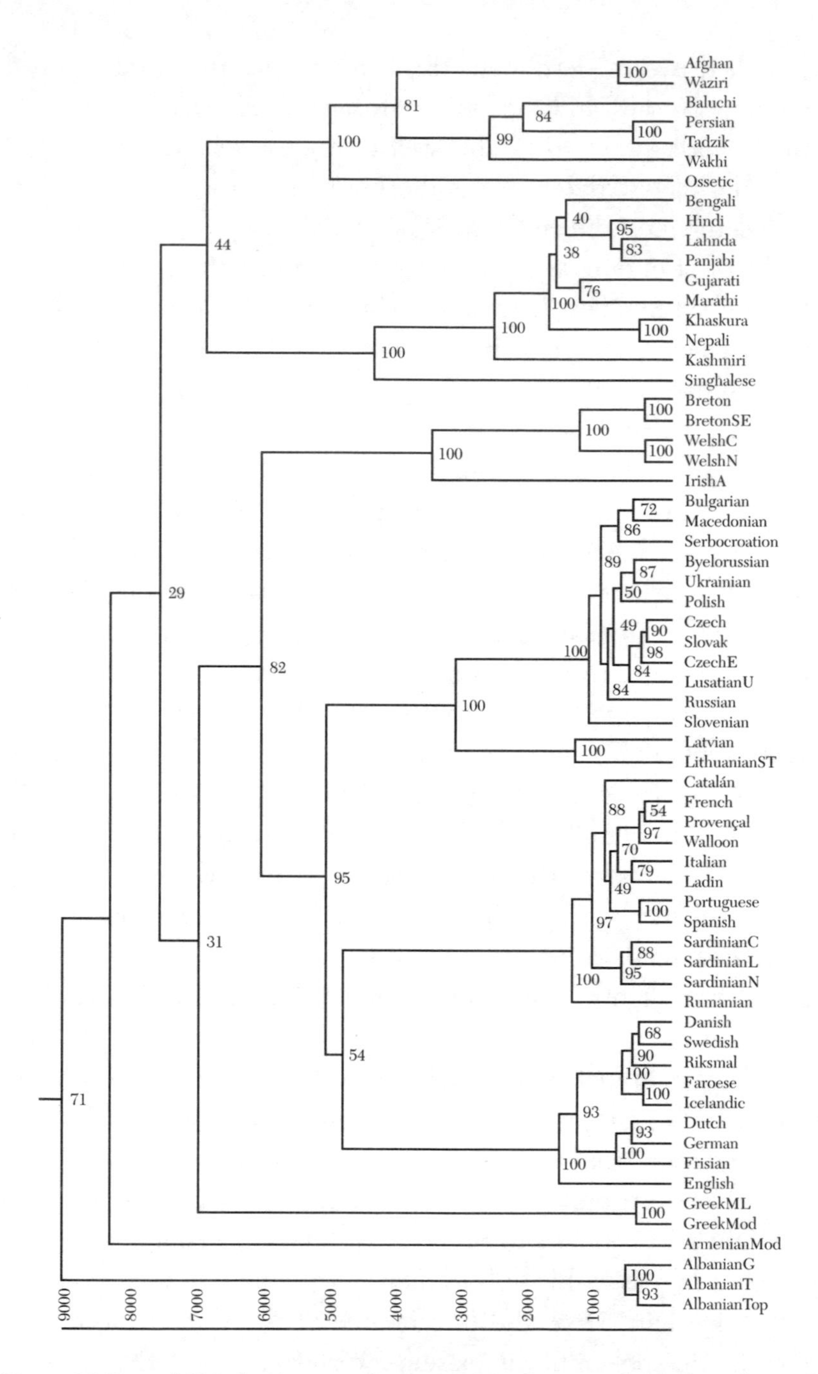

Figure 13. Tree of 63 Indo-European languages. Numbers near branches indicate the reliability in percent of the specific branch, calculated by the method of the bootstrap. The scale on the bottom indicates years. (From an unpublished manuscript by Piazza, Minch, and Cavalli-Sforza, based on data from Dyen, Kruskal, and Black 1992)

to completely replace the earlier pre–Indo-European language Basque but was successful in replacing Etruscan in the Italian peninsula. The Germanic branch settled in the northwest, north, and center of Europe, being unable to completely replace the earlier Indo-European subfamily which had also survived Celtic. The Baltic and Slavic branches settled in the northeast and the southeast, respectively, competing with the earlier settlers who spoke Uralic languages.

Recently, Tandy Warnow from Philadelphia and others have proposed a completely independent tree analysis of Indo-European history. Their results are not fully published and therefore difficult to analyze. They use a small number of word roots, believed to be especially reliable, and a very small number of languages that includes extinct ones. Their conclusions differ from ours mainly because they propose a very late branching of Celtic. This is difficult to reconcile with the early diffusion of Celtic languages to almost all of Europe, and their suppression by latecomers, speaking Germanic and Italian languages, which confined these languages to the extreme geographic periphery of northwestern Europe. The small number of words employed and the lack of a statistical test of robustness of conclusions may be the main problem of this otherwise very interesting analysis.

The Bantu Expansion

Many other expansions have brought new languages to new lands. The demic expansions of peoples that we are familiar with were almost all accompanied by their original languages. Among the prehistoric expansions studied both genetically and linguistically, the Bantu expansion is of considerable interest. Despite contacts and genetic flow from tribes speaking other languages, like Nilo-Saharan in East Africa, and the Khoisan in South Africa, the Bantus have largely maintained their genetic distinctiveness, which makes them somewhat different from the other West Africans from which they originated. Starting from Nigeria and Cameroon, they headed south, toward the Atlantic coast. The Bantus of the first expansion were still using Neolithic tools 3,000 years ago, but later Bantu expansion was aided by the introduction of iron. Only around A.D. 1 did the Bantus reach the Great Lakes region of Uganda and Kenya, and from there they expanded south, both near the coast of the Indian Ocean and further inland. From this time on, archeologists find Bantus relied heavily on iron.

The western and eastern currents moving into the south central continent met eventually. The Bantus were apparently still a few hundred kilometers from the Cape of Good Hope when the Dutch landed there around 1650. Both archeology and linguistics also show that the Bantus arrived earlier in Namibia along the West Coast.

Darwin's Prophecy

The origin of scientific linguistics can be dated to 1786. In that year, the English judge Sir William Jones advanced the theory that Sanskrit, Greek, and Latin, and possibly Celtic and Gothic (the ancestor of Germanic languages), appeared to have a common origin, at a famous conference at the Bengal Asiatic Society of Calcutta, which he had just founded and served as president. The similarity of Sanskrit and European languages had already been noticed by the Florentine merchant Filippo Sassetti and by the Jesuit priest Coerdoux. The latter sent notes from Pondicherry to the Academy of Inscriptions in Paris showing that Sanskrit, Greek, and Latin must have a common origin, but his conclusion did not have the impact of Jones's conference. In 1863 the German linguist August Schleicher published a tree showing the origins of Indo-Europeans very much like one we would draw today using modern methods. The ties between biology and linguistics were evident at once. Schleicher was certainly influenced by Charles Darwin's use of trees to explain the theory of organismal evolution. In *On the Ori-*

gin of Species, Darwin clearly stated that if we knew the tree of biological descent of the human groups, we could extract the tree relating languages. This effort was not attempted until 1988. I am ashamed to say I was not aware of Darwin's prophecy at the time. I was reminded by a friend who is a historian of our science and had read our 1988 paper, with Alberto Piazza, Paolo Menozzi, and Joanna Mountain, in which we correlated together on a global scale data of genetics, archeology, and linguistics. Here are Darwin's words:

> The natural system [of classification] is genealogical in its arrangement, like a pedigree. It may be worthwhile to illustrate this view of classification by taking the case of languages. If we possessed a perfect pedigree of mankind, a genealogical arrangement of the races of man would afford the best classification of the various languages now spoken throughout the world; and if all extinct languages, and all intermediate and slowly changing dialects, were to be included, such an arrangement would be the only possible one.

The correlation between genes and languages cannot be perfect, since the rapid conquest of large territories may favor replacement of indigenous languages with unrelated ones, as happened in much of the Americas. But these events do not appear to have occurred frequently enough to erase all trace of a correlation. We see equally that in the case of prolonged genetic exchanges with different neighbors, genes can be replaced. Nevertheless, despite the two sources of confusion, the correlation between genes and languages remains positive and statistically significant.

Even at a microgeographic level, the regions subject to detailed study have usually shown strong correlations between geography, genetics, linguistics, and other cultural aspects like surnames. Often the genetic-linguistic mosaic we observe clearly shows the effects of numerous expansions—some are known historically—and of their superimpositions and interactions. Perturbations do occur, but they do not manage in most cases to obscure the clarity of the correlation between genes, peoples, and languages.

A Hypothetical Model of Linguistic Evolutionary History Based Mostly on Genetic and Archeological Knowledge

Linguistic evolution is a subject of unusual interest. In this chapter, we have restricted ourselves to explaining the similarities between genes and languages. But linguistic evolution is also very important to understand as an example of a more general phenomenon, cultural evolution, which we will analyze in the next chapter.

Following Darwin's suggestion, one can attempt to use our knowledge of genetic evolution to make hypotheses about the earlier part of the linguistic tree. Figure 14 shows the linguistic history reconstructed when a number of cues taken from the genetic tree are correlated to linguistic information. Merritt Ruhlen drew the tree, using our 1988 genetic tree as a guideline. But he also took into account new linguistic superfamilies that had been daringly proposed in the interim. I have made few changes to his tree, adding some probable dates. When our genetic data is entirely satisfactory, the tree will probably not be as simple as this one, but it seems likely that its major features will not change.

The oldest linguistic families must be African: of the four existing today, Khoisan is considered to be the oldest; Afro-Asiatic the most recent; Niger-Kordofanian and Nilo-Saharan probably had a common origin (the Congo-Saharan superfamily suggested by some linguists), and must have arisen at an intermediate time. Khoisan may be the most direct descendant of the people who left Africa, as explained before.

For those Africans who remained, the first linguistic split is likely to have been between the branch leading to modern Khoisan on one side and a branch leading to the ancestral languages of Congo-Saharan on the other. Afroasiatic is likely to have originated much later, probably in northeast Africa, or otherwise in the Middle East, or in Arabia.

The Niger-Kordofanian family consists of two branches. The very small one is the Kordofan, named after a large mountain range in western Sudan, and the other is called Niger-Congo. Perhaps there was a westward spread from East Africa, first to the Kordofan,

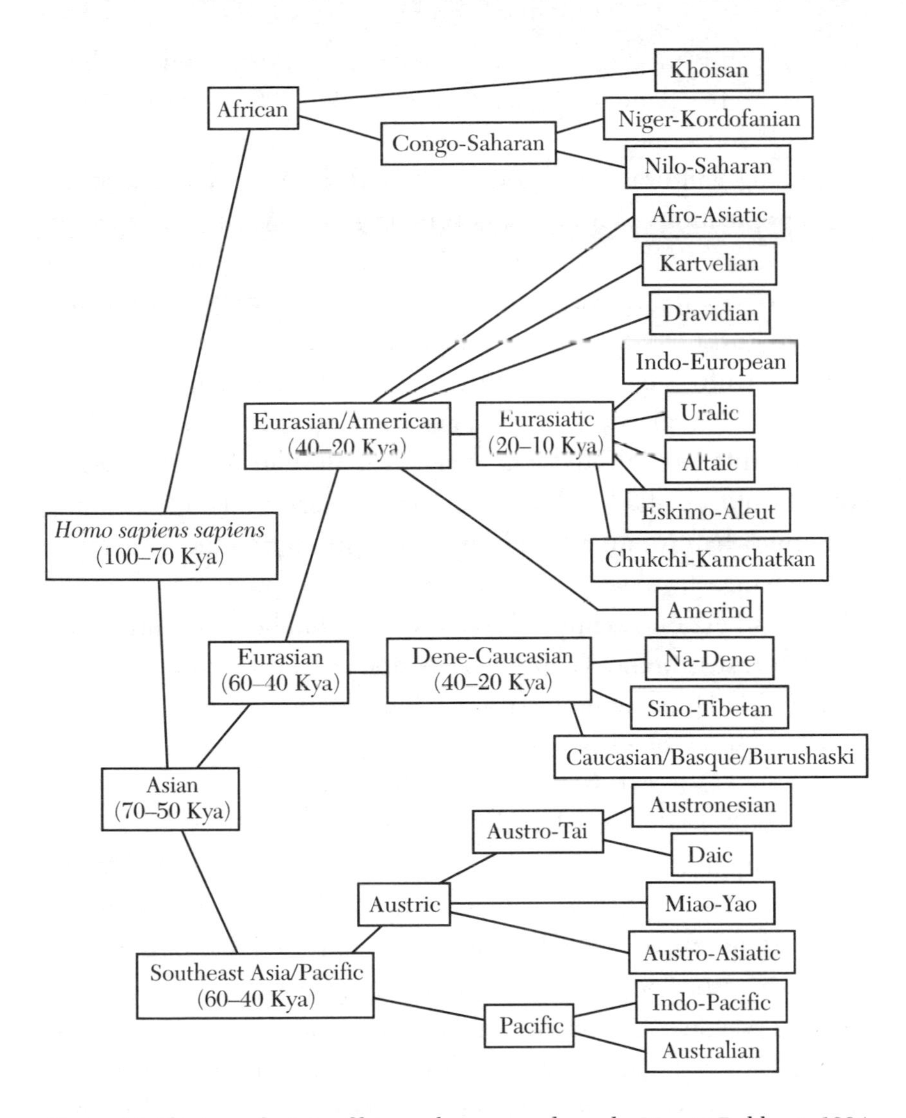

Figure 14. The tree of origin of human languages, drawn by Merritt Ruhlen in 1994, using Darwin's suggestion to base it on the genetic tree (given in figure 12), somewhat modified, with the likely range of initial divergence dates (Kya = 1,000 years ago).

then to West Africa, or the expansion may have been in the opposite direction. In West Africa a major population burst coupled with the introduction of agriculture may have occurred 4,000 to 6,000 years ago. I hope archeologists will note the genetic cues from principal component analysis of genes that suggest this may have begun between Mali and Burkina Faso (former Upper Volta). Agriculture then spread from Nigeria and Cameroon to central and southern Africa with the Bantu expansion, which took 3,000 years to reach as far south as it would.

The agricultural expansion originated in West Africa found a forest population of hunter-gatherers, Pygmies. Most of my work in Africa was in the forest areas of the tropical belt, where Pygmies have survived the arrival of farmers. They are very few and, wherever the forest has disappeared, their descendants are hardly recognizable. Unfortunately, the original Pygmy languages have all disappeared, replaced by languages of neighboring farmers. The only possible traces of these languages are in their names for forest animals and plants. Working in an area where foraging (hunting-gathering) barely survives side by side with agriculture has given me a chance to observe the transition from foraging to farming, which is still going on but will soon be over. I believe that archeologists interested in the Neolithic transition in Europe and elsewhere should spend time in this area, while it is not too late, in order to see a living model of what must have happened in similar circumstances in other parts of the world. All the farming people in proximity are dark-skinned, while those who live in the forest are less so.

A group of very dark-skinned Africans (sometimes called "elongated" because of their very elegant, tall and thin bodies), many of whom could make a good living as fashion models, inhabit East Africa and other neighboring regions. Linguistically, they are Nilo-Saharan, a name that indicates their dominant living area and their origin. The Nilo-Saharan started domesticating cattle in the Sahara perhaps 8,000 years ago, but many of them had to leave when it became a desert. Even today they are still mostly shepherds.

The earliest confirmed date of modern humans in Asia is in China, 67,000 years ago; and the first settlement of New Guinea and Oceania is believed to have been 50,000 to 60,000 years ago at the earliest, or at most somewhat later, 40,000 years ago. Did modern humans reach East Asia by land or perhaps by boat, along the southern coast? Probably both. There exist archeological estimates of the rate at which the recolonization of northern Europe took place at the end of the last glaciation (around 13,000 years ago): it varied from 0.5 to 2 kilometers per year, not too different from that of farmers. But the major limiting factor may have been largely that of ice withdrawal, rather than that of human movement. It is possible that the coastal route allowed a faster movement. How long did it take to travel by the hypothetical coastal route from East Africa, a likely point of departure, to Southeast Asia (a necessary point of arrival, from which some may have continued north to East Asia along the coast of the Pacific, and others south to New Guinea and Australia)? We may venture a minimum guess of 10,000 years. Perhaps modern adventurers might try to repeat a small part of the journey in conditions similar to those of our ancestors. This could be very informative, even if today's coast and the availability of seafood must be different from what it was then. Assuming that it took that much time to go from East Africa to Southeast Asia, the average displacement that may have taken place would have been on the order of 50 to 60 kilometers per generation (2 km/yr). This is about twice the rate of advance of farmers many thousands of years later. We are speaking of a mode of life of which there do not seem to exist living or historical examples (other than perhaps in Borneo). I would describe it as nomadic fishing. But the model of demographic expansion would not be very different from that of Neolithic farmers, in the sense that there must be both migration and reasonably active reproduction for an expansion to take place. Of course, over the generations, some extended families or small groups may have settled or chosen to leave the coast and go into the interior, as others continued their almost random wandering along the coast. The settlement of Southeast Asia, New Guinea, and Australia led to the development of the Indo-Pacific and Australian linguistic families. Andamanese and other Negritos of Southeast Asia are the closest living descendants of Africans who were among the first settlers of Southeast Asia and Oceania.

China and Japan may have been settled before Australia, and may have been the first areas of development of the Dene-Sino-Caucasian family, which must have spread west through central Asia to Europe. The Na-Dene branch went to Siberia, from which it later (about 10,000 years ago) migrated to North America, after the Amerindians first colonized it (some 15,000 to 30,000 years ago).

A late expansion centered in Southeast Asia was that of the Austric superfamily, which connects genetically Taiwanese aborigines and the southern Mongols of Southeast Asia. The second PC of Asia indicates a possible genetic expansion centered in Southeast Asia. This may have happened both very early, and also again as late as the local development of agriculture. Principal components cannot distinguish two expansions having the same area of origin which took place at different times.

The multiplicity of migrations and expansions between Europe and Asia, in both directions, is well substantiated. A recent discovery of a late west-to-east migration was that of the Indo-Europeans who went as far as western China as recently as 4,000 to 1,000 years ago, but the languages spoken by them, Tocharian, are extinct. Among the last people to expand from a center located in the east were Mongolians who engaged China, forcing the emperors to build the Great Wall more than 2,200 years ago. Attila the Hun went as far west as Italy. Their relatives speaking Turkic languages started expanding from Central Asia about eight or nine centuries ago, eventually reaching Turkey and the Balkans.

We have mentioned that the roughly continuous genetic gradient from Europe to East Asia is the result of all these migrations. Many Central Asians are herders and nomadic. Languages, especially those forming the Eurasiatic family, introduce apparent discontinuities in the gradient, as communities inevitably speak a single, common language. Some extend widely and may move, creating a complex human geography. Political changes and military events may force the replacement of languages in a relatively short time. Here the correlation of languages and genes cannot be perfect but it is still evident to some extent, in spite of the turbulence of the Eurasian history of the last four or five millennia.

Genetic research can certainly help the understanding of linguistic evolution, and vice versa.

CHAPTER 6

Cultural Transmission and Evolution

Humans differ from other animals—even their closest cousins—by the richness of their culture and the importance it is accorded. Culture, liberally defined, is not restricted to us and can be observed in other species. Anthropologists have proposed literally hundreds of definitions of culture, which are mostly abstract and exclude technology. I prefer the converse, and give the simplest and broadest possible definition: culture is the ensemble of customs and technologies that played and continue to play an essential role in the evolution of our behavior. Such a definition includes animal cultures, although they are less developed than the human one because animal communication is clearly much more limited. But one must add to the definition that culture is what is learned from others, especially our ancestors. What we add to it by our own efforts are usually modest innovative contributions generated by independent, solitary learning. These are sometimes passed to others and thus become part of culture, as they become available to future generations. The cultural track is the only one that allows knowledge about the world to pile up over generations. It removes the limit of a single lifespan on the accumulation of information.

Teaching provided by parents (especially the mother) is essential for most bird and mammal species. There are also other forms of indirect teaching as well as innate behavior, exemplified by imprinting in birds: fledglings are geared by nature to identify the individual with which they pass their first hours of life outside their shell as their mother, as well as the species to which they belong. Depending on the species of bird, the process can be more or less complicated. Imprinting is a form of biological adaptation probably found in humans as well, although it remains poorly studied. It can be identified by the existence of a "receptive period" or a "critical period" for a specific learning.

Human education occurs primarily by imitation or direct teaching (oral or written). We usually don't make a formal distinction between these two mechanisms of human education. There is always at least one transmitter and one receiver, and information that passes between the two. Language greatly increases the efficiency of this process and forms the very basis of human culture. More than anything else, it has allowed humans to adapt to and master their surroundings in a very short time. Throughout human evolution, language has given modern humans much of their advantage over other species and made possible the complexity of our knowledge today.

Language is an innovation that involves both biology and culture. It is the result of natural selection acting on anatomy and physiology. Children are born with the propensity and ability to learn a language. It is even likely that Neandertals had similar but less developed language abilities. (It has been said that the Neandertal larynx was not long enough to produce our richness of vowels, but there is not yet enough evidence to support such a statement.) While language itself is a cultural creation, it requires a precise anatomical and neurological foundation. This development probably came very gradually and progressively. *Homo habilis* may have been able to speak in some fashion even two million years ago. Phillip Tobias noticed a larger cavity next to the left cerebral hemisphere in six *habilis* skulls analyzed. It is here that a cerebral protuberance exists which is known to be one of the neurological centers for speech—Broca's area. Tobias's observation suggests that this center had already achieved a certain degree of development in the first species that we place in the genus *Homo.* A similar bulge is not found in monkeys.

Culture as a Means of Biological Adaptation

The ability to learn is one of the most fundamental characteristics of life—even in very simple organisms. Culture, or the ability to learn from the experience of others, is a special phenomenon that relies on communication. The speed and the precision of communication, and even our ability to memorize what we learn, are factors that govern the efficiency of culture. Naturally, it is not enough that culture exists for it to be useful from a biological viewpoint. But several examples can demonstrate its potential value for biological adaptation. On their own, our senses of taste and smell are not enough to help us to safely choose the food we eat; we must also learn from someone else to recognize which plants are toxic and which animals are dangerous.

Culture enables us to accumulate prior discoveries and helps us profit from experience transmitted by our ancestors—knowledge that we would not have on our own. In principle, it has always been possible for a lone individual to invent differential and integral calculus starting from scratch, but the odds are very low. Even Gottfried Leibniz and Isaac Newton used existing mathematical knowledge in making these fundamental contributions. Until the invention of writing, the accumulation of knowledge was limited by human memory, which varies from one person to another. Today, this limit has disappeared. The abundance of information in the last twenty years is changing the world thanks to the rapid access modern communications provides to it. Such change was unimaginable even a few years ago.

Culture resembles the genome in the sense that each one accumulates useful information from generation to generation. The genome increases adaptation to the world by the automatic choice of fitter genetic types under natural selection, while cultural information accumulates in a person's nerve cells, being received from another person and selectively retained. Cultural transmission occurs in a variety of ways: by the traditional path (observation, teaching, conversation), through books, computers, or other media developed by modern technology.

Evolution also results from the accumulation of new information. In the case of a biological mutation, new information is provided by an error of genetic transmission (i.e., a change in the DNA during its transmission from parent to child). Genetic mutations are spontaneous, chance changes, which are rarely beneficial, and more often have no effect, or a deleterious one. Natural selection makes it possible to accept the good ones and eliminate the bad ones. Cultural "mutations" can be accidental and minor like many genetic mutations—mistakes in the copying of manuscripts in medieval monasteries, for example. Minor variation would result from the errors introduced by a scribe in copying a manuscript. Most of these errors are probably accidental, resulting from inattention. And sometimes, the scribe will take the initiative and make a change that, in his opinion, helps comprehension or the quality of the text, but that may confound future philologists.

There is a fundamental difference between biological and cultural mutation. Cultural "mutations" may result from random events, and thus be very similar to genetic mutations, but cultural changes are more often intentional or directed toward a very specific goal, while biological mutations are blind to their potential benefit. At the level of mutation, cultural evolution can be directed while genetic change cannot.

But we inevitably arrive at the impression that most innovations are rarely truly advantageous. Sometimes the person suggesting an innovation makes a profit from it, but innovations that should improve the state of an individual, or of a social situation, often miss their mark and turn out to be unimportant, inappropriate, or even disastrous. Political history is full of examples. One of the most common errors is the exaggerated confidence in the heritability of political skill; the son of a powerful leader frequently is appointed to follow in his father's footsteps. The effects are often very disappointing. Mendelian inheritance predicts this problem, because the similarity between parent and offspring is on average modest. History shows that hereditary monarchies last only for a short time. When stripped of genuine authority, they are often incapable of appropriately performing even their symbolic roles. Nevertheless, selection generally tends to create and maintain customs and institutions with social utility. Even if imperfect or detrimental, some cultural changes are adopted and persist, occasionally incorporating modifications based on experience. The continuous changing of customs makes us forget the original purpose of a particular practice; without history, it quickly becomes difficult to reconstruct the reasons for certain rules and social conventions. One example that deserves further research is reproductive control in economically primitive cultures, which appears to have been quite common for a long time before its post-Paleolithic decline. Then, as now among the Pygmies and perhaps all modern hunter-gatherers, pacing births helps slow population growth to manageable rates, avoiding disastrous population explosions. It was only during the Neolithic—or in general with the development of agriculture—that populations began to grow rapidly, since more people could be fed in agricultural societies. Pygmies do not like to have children more than once every four years and believe that conceiving a second child too soon after another places the first at great risk. I doubt that the Pygmies consciously realize that this provides an important restraint on population growth, and they generally offer other explanations for the custom. Demographic stasis is usually important and necessary for peaceful cohabitation of different people, but so is the ability for nomadic populations to move without the burden of carrying several small children. With an interval of four years between births only one child needs to be carried by a parent, and the population remains stationary or grows very slowly. Maintaining this four-year gap between children requires great discipline. Some researchers think that breastfeeding—by preventing ovulation—can also prevent a new pregnancy, but that does not appear to suffice. The truth is that Pygmies avoid frequent pregnancies by observing a sexual taboo for three years following the birth of a child. They make this sacrifice for the health of their children without thinking of the long-term advantages that result from this celibate period, which on its own is not likely to provide sufficient motivation. It seems to me that this sexual taboo would disappear if breastfeeding for three years were itself sufficient to prevent pregnancies. The conclusion to draw is that reasonable reproductive customs arose during the Paleolithic among hunter-gatherers that have helped them keep their demographic growth near zero, probably without their knowing or realizing it.

Every day, we face choices that may be trivial or may affect us for years. These choices are a sort of "cultural selection." Unlike natural selection, which chooses between the best naturally adapted individuals of a species, cultural selection proceeds through the choices made by individuals. Ultimately natural selection will operate, since it works on the cultural choices we make as well. If our choices help us reach maturity and reproduce, then our cultural decisions (as well as biological predispositions) that generated these particular choices will be favored by natural selection. Thus, each cultural decision must pass two levels of control: cultural selection acts first through choices made by individuals, followed by natural selection, which automatically evaluates these decisions based on their effects on our survival and reproduction. Every cultural decision will also be favored by natural selection if it affects survival and reproduction, creating a positive correlation between these two forms of selection.

Although culture can intervene and modify them, innate impulses were passed down to us from our ancestors, upon whom natural selection acted. Often quite strong, these impulses are rarely absent. Many of our sensations and actions are either pleasing or painful, and frequently determine our behavior. It requires a certain amount of reflection to identify these impulses, but they can be appreciated by noticing the emotional charge potent in certain words. We can also observe when a word picks up an emotional overtone from its context: "drunk with happiness," "drunk with power," and "drunk with sorrow."

This emotional coloring is certainly the result of cerebral structures, although they are poorly known. We know some centers of the brain that, when artificially stimulated, provoke sensations of pleasure or pain. These internal centers, called "reward centers," undoubtedly influence our decisions, although a higher level of decision making must exist since we can also make decisions that we know will cause pain. For this to happen, there must be an alternative, more important motivation that makes us accept costly or painful decisions. In any case, it is clear that pleasure, pain, and sorrow, or the expectation that one or more may occur at some future time, can influence our decisions. It is at this level that we can most easily see the possible dissociation of cultural from natural selection. Drugs that induce pleasure carry the risk of death or disability. The conflict between sexual desire and a knowledge of the danger of AIDS or other sexually transmitted diseases offers another contemporary example. The fear of hearing an unwelcome diagnosis can keep us away from a doctor who might be able to help. In the Fore tribe of New Guinea, relatives customarily ate their dead. When an infectious disease—kuru, most probably related to Mad Cow Disease—struck the Fore, extraordinary persuasion was needed to discontinue this form of cannibalism, considered a duty toward their ancestors, which spread the disease. Biological and cultural tendencies often come into conflict, and, in order to avoid harm, we must not obey every innate—or even every learned—impulse.

How Is Culture Transmitted?

We acquire our culture from those around us and pass it on, in turn, to others. An important distinction between modes of cultural transmission must be made. We have borrowed terms from epidemiology to describe the two principal routes of transmission: vertical transmission describes the passage of information from parent to child, while horizontal transmission includes all other pathways between unrelated individuals. Evolution is slow under vertical transmission, which resembles genetic inheritance, because the time unit is the passage of a generation. Horizontal transmission, however, can occur rapidly—sometimes resembling an epidemic disease spread by direct contact between a susceptible and a contagious individual.

The ability to modulate the rate of evolution makes culture a very powerful agent of change. Special types of cultural transmission can profoundly affect the rates of change. For instance, when an idea is simultaneously spread from one to many people, very rapid evolution may result. Change comes more slowly when horizontal transmission is only from person to person (e.g., by word of mouth). Intermediate in rate is transmission by a hierarchical route. These distinctions produce important differences in the dynamics and success of cultural change. We have examined two aspects in particular: variation of a characteristic over time and variation among members of the same social group and among social groups. In collaboration with Marcus Feldman at Stanford, I studied the theoretical consequences of different mechanisms of cultural transmission.

Cultural transmission occurs necessarily in two steps: an idea must first be communicated and then it must be accepted. Any communication can be misunderstood, forgotten, or simply be made in an unconvincing way. In general, no innovation is assured of success. Often, something must be repeated to meet a favorable reception. If the originator possesses unusual charisma, prestige, or political or religious authority, the likelihood of successful acceptance is increased. The ages of the communicator and of the receiver are also important. In the theory explained below, we only consider transmission to have occurred in those cases where a cultural innovation is accepted by its receiver.

We can identify several forms of vertical transmission and three types of horizontal transmission comprising a transmitter and a single receiver, a single transmitter and several receivers, or several transmitters and a single receiver.

1. Vertical transmission occurs between a member of one generation and a member of the following generation. A biological rela-

tionship between the two sides is not necessary, since an adopted child may be equally receptive. The extent of parental influence—whether the child belongs biologically or by adoption—is usually great. This form of transmission has evolutionary consequences that are very similar to biological transmission, especially if the transmission occurs via only one biological parent, or an adoptive or cultural "parent" (uni-parental transmission): the rules are almost the same as the simple rules of genetic inheritance (of mtDNA, or Y chromosome, for example). Vertical cultural transmission can be just as conservative as genetic inheritance. Variation is introduced only through cultural mutation or the immigration of individuals coming from another society who have something new to teach. Transmission from grandparents to grandchildren is more conservative still—by a factor of two—and transmission over several generations can maintain important cultural features over long periods of time. Vertical transmission was certainly enhanced by writing. Examples are the influence of Greek philosophers like Plato and Aristotle; or the patriarchs of the Catholic Church like Saints Augustine and Thomas Aquinas. The oral transmission of religious texts before they were committed to writing also allowed rigid conservation, however, including that of rites and dogma.

2. Horizontal transmission, resembling the spread of an infectious disease, occurs between two individuals of the same or different generations who do not have the clear-cut biological or social relationship recognizable in vertical transmission. In epidemics, the contact that transmits a disease between two people can be very brief; but cultural transmission usually requires more prolonged contact. When the transmitter belongs to an older generation than the receiver but is not a parent, we speak of "oblique transmission." It insures that information passes from one generation to another. A more complete analysis of a population's structure by age and probability of transmission according to the age of the transmitters and receivers is possible, but the mathematics involved are often prohibitive.

The theoretical problems of horizontal transmission are similar to those facing the study of epidemics of infectious diseases, which have been studied in great detail. These analyses can actually be applied almost directly to horizontal cultural transmission. In effect, a successful cultural mutation sets off a cultural epidemic. The number of people converted to the new cultural characteristic increases in time according to a "logistic" curve. This curve has a maximum rate of increase at the beginning, slows down to a constant rate of increase, which lasts for an extended period of time, and eventually flattens until it reaches a maximum that may include the entire population or only a fraction of it. Geographic, social, or economic barriers impose the primary limits to cultural diffusion. The chance of success is determined by a number of factors, beginning with the attractiveness of the idea to its potential converts. The question to answer is whether the success of a cultural novelty is similar to an infectious disease where the ability of the parasite or virus to reproduce in its host must exceed a certain calculable threshold for an epidemic to be sustained.

3. Judging from ethnographic data, social structure has become considerably more complex than it was in the relatively egalitarian society of hunter-gatherers that followed the development of agriculture. With the increased size of social groups, the authority provided by chiefs and clans became necessary. Society became stratified into social classes, often within a defined hierarchy. Under these conditions, the transmission of a chief's will to all of a group's members made it easier for innovations to be passed from one individual to many others. A similar type of multiple transmission also arose when education and teaching were formalized, and a master had several students. The speed and efficiency of transmission from one to many is reaching its theoretical limit in the modern media. Information about important events can be communicated simultaneously to a billion or more people. In our information age the greater number of role models among whom we can choose and accept voluntarily can have extraordinary influence.

Cultural transmission is easier, faster, and more efficient when a powerful, authoritarian chief forces the acceptance of an innovation. Many societal changes are probably the result of the will of a powerful or a charismatic authority. Popes have the ability to propose new dogma, which must be accepted by the faithful under penalty of excommunication from the Church. In a less serious vein, the Fascist government in Italy tried to influence language use by declaring war on French and English words that had crept into the language. It also wanted to suppress the use of the third person singular pronoun *lei* and generalize that of the second person plural *voi* for respectfully addressing people. The usage of the third person singular derives from the Spanish *usted.* It was imposed on Southern Italy by a monarchy of Spanish extraction after the Aragonese conquest. But the Fascist attempt to abolish the use of *lei* and other foreign words failed, though an invented word—*autista* (driver)—replaced a word then in common use, of French origin, "chauffeur," which is difficult to pronounce in Italian. It is not easy to force things on Italians. Fascism's greatest success was perhaps to compel party membership and to force adult males to wear party symbols. Compliance was assured by making it a necessary condition for almost every type of employment.

More important are some cultural changes that spread throughout Tibet and parts of India. Polygyny (the practice of keeping multiple wives) and polyandry (the practice of keeping multiple husbands) became popular practices, which still exist. These two forms of polygamy are sometimes found in the same village; there are even simultaneous marriages between multiple men and women. The wives and husbands in these multiple marriages are usually siblings, which probably explains these arrangements, since they avoid dividing inheritances and agricultural lands between siblings. This is a very bold solution to a common problem, which is acute in the marginal agricultural environment of Tibet. Elsewhere, this problem has been resolved, perhaps unfairly, through primogeniture, under which the eldest child (or son) inherits all property. The history of polyandry in Tibet is not well known—some record of it may be preserved by monks or monastic documents. One possible hypothesis suggests that feudal lords, with the consent of religious leaders, were allowed to experiment with and implement these social changes, which may seem extreme to us today. I should admit, however, that my wife's three uncles—the three Buzzati brothers, of whom Dino was a well-known author in Europe, and Adriano my genetics teacher—declared that they all wished to marry the same woman. Two of them never did marry, perhaps because this plan could not be implemented.

4. The inverse mechanism—several transmitters to one receiver—is also very important. In a social group, several members (and sometimes all) often exert psychological pressure on new members. Each latecomer can thus find himself the object of strong pressure from many sides, in what is often a more persuasive procedure than when a single transmitter is influence. This "social pressure" can even occur in small groups, like nuclear families. The mechanism of transmission by multiple transmitters, usually acting in unison, has been called "concerted." It tends to suppress individual variation and to homogenize a social group. It is thus the most conservative mechanism of all.

The family is the most important social group, exercising considerable pressure over its members—particularly the youngest ones, who may not have developed critical judgment or the ability to resist. But we all know that some people can resist most influences. Rebellion tends to develop only later, but when this mechanism of social pressure meets little resistance, it is the most powerful.

Hervé Le Bras and Emmanuel Todd (1981) have recently refined ideas by the French sociologist Fredericq Le Play. They believe three major types of families exist in France: (1) The family with absolute patriarchal authority in the northwest in which the head of the household makes all decisions on behalf of the family's members, a custom that may have been inherited from the Celts. (2) A more relaxed form of patriarchy that emphasizes mutual support and allows the children to marry, have children, and continue to live in the family home if they are unable to support themselves. Older members also live with the family and are cared for by relatives. This type of family is common in the southwest in an area that corresponds, at least according to genetic data, to the proto-Basque area. (3) The strictly nuclear family frequent in the northeast, in which children can marry and have children only if they have the ability to live independently. This practice is most frequent where the Franks had the greatest influence. Franks were barbarians of

Germanic origin who conquered France in the early Middle Ages and later extended their control to the rest of France. It is interesting to note that recent historical research has shown that this type of family was also common in Germany and in England after the Anglo-Saxon conquest. This arrangement that, among other things, encourages youth to relocate in the search for employment probably favored industrial development.

Le Bras and Todd have proposed a controversial but stimulating hypothesis that says family structure influences political outlook: customs learned in the family microcosm largely determine those that will be more easily accepted when the young are introduced to the social macrocosm. Family members search for social systems that mimic, to some extent, the family life with which they are familiar. That is why monarchy and authoritarian systems may be more popular in northwest France than in the southwest, where the Socialist vote is quite strong; in the northeast, the vote favors a free market economy. Todd (1990) applied this analysis with success to other parts of the world. It is also interesting to note that the division of France by family type shows a strong correlation with genetic history. I don't believe it is worth looking for a genetic explanation here; the sociological explanation for the agreement between the family microcosm and the social macrocosm seems consistent with our theory of cultural transmission. The correlation of sociological differences with genetic differences is simply the consequence of ethnic separation. Deep and ancient ethnic differences can easily be conserved for twenty centuries or more, thanks to the conservation of family structure. This conservation itself is due to the inevitable fact that family structure is inherited by vertical transmission and reinforced by powerful social pressure within the social group, which acts on new members when they are young and most sensitive.

This hypothesis was confirmed by an independent study of Rosalba Guglielmino and others (1995). In an analysis of data from Murdock's Ethnographic Atlas, restricted, for the time being, to Africa, we have observed that the most conserved cultural characteristics are familial. This study showed that very few other cultural characteristics are as highly conserved although some cultural traits, like the form and structure of dwellings, as well as certain socioeconomic characteristics that depend on the degree of social evolution, do not change rapidly.

With Marcus Feldman, our team has studied the evolutionary consequences of these transmission mechanisms—the way in which social groups evolve—when a social innovation is introduced (1982). Will it establish itself readily or not? We chose to do this work mathematically, which has the advantage of precision but is not always appreciated by readers. It is perhaps for this reason that anthropologists have not shown much interest in these models, unlike economists, for example, for whom the use of mathematics poses no problem. However, one could reach the same conclusions by using just a bit of common sense. And it is worth reiterating that analysis of cultural transmission, and in particular the distinction between vertical and horizontal transmission and their major forms, is essential to our understanding of cultural inheritance.

A novel behavior could be a variant of an existing custom that has not been universally accepted. It could also be a completely new invention. When parents teach a new behavior to their children, there is a very good chance that it will be accepted, since children are more receptive to new ideas than adults. Acceptance within the family may be successful, but, like genetic inheritance, many generations of transmitting the innovation, or other mechanisms of transmission, are required to spread the cultural change from one family to many, or to all members of a society.

In the case of horizontal transmission, the diffusion of an innovation through a population can be more rapid (occurring within the space of a generation, in some cases) as long as learning it is easy and its consequences are agreeable. As with an epidemic, diffusion can stop before reaching the entire population.

The speed of innovation adoption is maximal when one person communicates with many others. The decision of an authoritative political chief will be accepted almost immediately by all subjects, as long as it does not present serious disadvantages. History shows that many social and political events were completely determined by monarchs or by influential people of the ruling elite. In modern society democratic principles have established more complex political processes, although a small number of people from the political or business world continue to control an important number of daily decisions. The hierarchical structure of society can help a transition proceed, starting with the upper echelons of power and descending to the lower ones.

On the other hand, under the fourth transmission mechanism, which we have called "concerted" (where many people transmit the same cultural trait to a single person, usually of the next generation, and do so in turn with each member of it), an innovation has a very small chance of success. A single person sympathetic to the change would first meet resistance among the allies he needs. Unless the innovation is unusually useful, or the proposer prestigious, success is unlikely.

Most cultural characteristics are transmitted by a variety of means that often conflict. This conflict is common; for instance it occurs when one learns rules of behavior at school that differ from those taught at home, or when companions at school have very different opinions from school authorities and/or the family. The consequences of these conflicts vary significantly among individuals and particular cultural traits.

Examples of Cultural Transmission

Cultural transmission comprises education received from both family and school. It also includes all the habits and customs that are not explicitly part of one's education. Certainly, an individual acquires these through personal experience, but here again, conscious or unconscious imitation must play an important part.

It is not easy to distinguish between relative contributions. The similarity between two friends or between two individuals with a more intimate relationship, like a husband and wife who have lived together for a very long period of time, are partly the expression of what the two have learned from each other and what may have attracted them in the first place. These forces are often very strong, and we sought to examine them by surveying a group of students about similarities between husbands and wives, parents and children, and between friends. We asked about forty questions, and queried the students about themselves and their parents, as well as the parents about the students and themselves. On average, the correlations (the similarities) between husbands and wives (the students' parents) were the greatest, followed by those between parents and children, and finally between friends. The characteristics studied addressed social activities, habits, leisure activities, superstitions, beliefs, and so on.

The most interesting result of this study is that the highest correlations were shown by characteristics of two categories—religion and politics. Both show the major role played by parents, that is, by vertical cultural transmission. In the first case, children resembled their mother to a remarkable degree, both in the choice of religion in mixed marriages, and in the frequency of prayer. The choice of religion is not surprising, since a child's religion is almost always chosen by his parents, or at least by one of them, at an age when the child cannot express his own preferences. Conversions do occur, but only rarely, and later in life. That a twenty-year-old should continue to pray to God, however, does seem to imply more than familial constraint. Unfortunately our data do not indicate whether prayer continues to be an important activity throughout the life of children raised to pray. If the mother's influence prevails in the choice of religion, the father appears to exercise influence only on the regularity with which a religion is practiced, which is a social rather than a spiritual decision—and even in this case, the mother's influence is as great as the father's. Both parents appear to contribute equally to a child's political outlook.

It is always possible that some part of the similarity between parents and children has a genetic basis. The distinction between biological and cultural transmission is not always an obvious one. For example, it was long believed that the similarity between the IQ of parents and children was entirely genetic in origin. The famous British psychologist Sir Cyril Burt, undoubtedly carried away by enthusiasm, even stooped so low as to publish false data to "prove" a genetic basis for IQ. It is thanks to the American psychologist Leon Kamin that Burt's fraud was discovered.

At the beginning of work on IQ, the French government asked Alfred Binet to develop a method for identifying children with mental handicaps in order to provide them with special schooling at an early age. But it was primarily American psychologists who tried to change Binet's IQ scores into a measure of "pure" intelligence independent of the culture or social milieu in which the tests were conducted. This misplaced enthusiasm led to several serious social mistakes, of which not all have been corrected. The study of adopted children was decisive in showing that cultural transmission exerts a strong influence on the determination of IQ. American studies in 1980 and 1981 established that only one-third of the variation in IQ among individuals was due to biological heredity. Another third can be explained by cultural transmission, while the last third appears mostly due to other unspecified, mostly random differences in personal life experience. This is a far cry from the 80 or 90 percent genetic contribution suggested by Burt and his many American colleagues. Similarly, Arthur Jensen's statement that the low IQ average of African Americans relative to Whites is genetic was contradicted by studies of Black children adopted by Whites in England and the United States.

Theories about the role IQ plays in social stratification have also been disproved. Some researchers have claimed, without real evidence, that the difference in IQ observed between high and low social classes was genetic, because people with a high IQ automatically became part of the high social classes. A French study on adoptions again showed that the difference was primarily sociocultural and not genetic.

There is probably still very widespread prejudice in America concerning the low IQ of Black Americans: the majority is likely to be still convinced that it is the result of a real genetic difference and not of a strong social handicap that cannot be reversed in a short time. Contrast the enthusiastic acceptance of the book *The Bell Curve* and its racist message with the response to the information that the average Japanese IQ is greater than that of White Americans by 11 points, almost as much as the average difference between White Americans and Black Americans. Then, the response was: it is clear that American high schools are very bad.

Adoption studies provide the best guarantee against the confusion of biological and cultural transmissions, but these studies are difficult and costly, largely because there are so few subjects. The most ambitious studies use identical twins who have been raised separately. But these studies are hampered by small samples and because the very early environments of twins, pairs, and their rearing is not always independent. But other methods exist that help limit confusion between cultural and biological inheritance. For example, in the case of religious or political similarities between parents and children, we used published research data comparing identical twins, fraternal twins, and regular siblings. Fraternal twins should not resemble each other more than regular siblings if biological heredity were the only important factor. In the case of religious or political creeds, the similarity between fraternal pairs was almost the same as for identical twins, indicating that genetics plays no (or only a very small part) in this trait. Family background does have a major effect. The purely or predominantly maternal transmission of some religious characteristics would be difficult to explain in a strictly biological way. Maternal transmission exists for biological characteristics determined by the mitochondrial genome. However, it would be very surprising if these cytoplasmic organelles, which supply the cell's energy, had any effect on individual religious beliefs.

We can study cultural transmission directly, instead of taking the indirect approach of twin studies. In this way we avoid confusing biological inheritance with other mechanisms. We can question people directly for certain characteristics, and the depth of memory revealed in subjects is often surprising. In collaboration, the anthropologist Barry Hewlett and I asked African Pygmies from whom they learned certain basic knowledge essential to their life: information about hunting, gathering, preparing food, building houses, and so on. They perfectly remembered learning these things and often they even remembered the time and place where they were taught particular skills. The information collected could be verified by questioning also the teachers of this knowledge. From 80 to 90 percent of the time, parents were responsible for the transmission. Because some of the skills were known to only one sex, teaching usually occurred via the parent of the same sex. Only for important communal activities like dancing, singing, rules for dividing food, and other characteristics of Pygmy society did members of the wider community participate with the parents in educating children. The contribution of other African villagers with whom Pygmies associate during parts of the year is restricted mostly to farming, which Pygmies do in a very limited way, having been exclusively hunter-gatherers until recently. The Pygmies have also learned how to make and use certain hunting weapons, like the crossbow, from villagers. This knowledge has diffused very quickly among the Pygmies. Our notes show that several Pygmies learned how to make crossbows directly from a villager, although we also observed one instance of a Pygmy father teaching his son. A traditional society like that of the African Pygmies that lacks leaders and schools, and is organized in very small social groups, tends to remain independent from nearby villagers, even when those villagers try to establish control over the Pygmies. Therefore, cultural transmission tends to be vertical within both groups, with very limited horizontal exchange between them. Vertical transmission and the social pressure exerted by the group's members tend to make Pygmy society very conservative. African farmers, on the other hand, have more contact with the exterior, missionaries, for instance, and have radios and schools.

Critical (Receptive/Susceptible) Periods and Their Importance

Most culturally determined characteristics are more easily changed than genetic ones. Even for clear-cut genetic diseases, onset can occur very late in life, with much variation from one individual to another. Huntington's chorea can strike individuals ranging from two to eighty years old, although most cases manifest themselves around the age of forty. But the pattern of inheritance is very strict. Some genetic diseases disappear with age, as do certain types of allergy. But in general, genetically determined characteristics are rather stable and rarely reversible. The same is not true of cultural characteristics. We have already observed that religious conversions take place. Political affiliations also change with appreciable frequency.

Nevertheless, some cultural traits change less readily than others. Stability of certain behaviors may be favored by biological factors that render changes more or less likely at certain ages. In other words, there may be sensitive or critical periods in life, the phenomenon sometimes called "imprinting."

The most obvious critical period, although inadequately studied, is undoubtedly the one governing our ability to learn a first and a second language. The first language must be acquired in the first years of life. One can learn other languages after the first, but rarely, if ever, as well; it is particularly difficult to learn proper pronunciation of a foreign language after puberty.

The time before puberty is also a sensitive period for acquiring the incest taboo. The psychologist Edward Westermarck has suggested that the cohabitation of brothers and sisters before puberty could diminish sexual interest and explain why incest between brothers and sisters is as rare for humans as it is for other mammals. There have been notable exceptions in some ancient dynasties such as those in Egypt and Persia, where the marriage of siblings was encouraged, but this custom quickly disappeared. In some communities, especially in the Middle East and India, marriages between close relatives (e.g., uncle and niece, first cousins) are still frequent, but this is a different phenomenon.

Westermarck's hypothesis was tested by Arthur Wolf (1980) in Taiwan, where marriages have occurred between a boy and an adopted sister of similar age. The daughter would be adopted by the boy's parents after his birth. In a society where spouses are bought, adoption at a very young age guaranteed a lower buying price. This custom also afforded the mother the opportunity to instruct the future daughter-in-law in the art of serving her husband. Wolf showed that these marriages were less successful than others; they ended more often in divorce and produced fewer children on average. This result is consistent with data from Israeli kibbutzim where children are raised together in a sort of communal nursery, and have little contact with their parents. These children essentially have a very large family of adoptive brothers and sisters, and very few marriages occur between children from the same kibbutz. It is harder to fall in love with someone whom you are used to seeing on the potty.

There are certainly many other critical periods in the formation of human societies, about which we don't currently know much. Even those that I've just cited have not been studied in sufficient detail. I could mention just two other areas deserving further inquiry. Gianna Zei, Paola Astolfi, and Suresh Jayakar have shown that daughters of an older father tend to marry men considerably older than themselves. This may be part of a more general phenomenon, which deserves to be investigated in great detail: it is likely that we have a tendency to choose spouses who share some physical resemblance (and perhaps behavior) with our opposite-sex parents. This phenomenon could explain the pronounced physical resemblance observed among individuals of the same social group—especially obvious in the small and isolated ones. The same phenomenon broadens differences between groups.

In another investigation aided by psychologists, we studied the propensity of our Stanford students to identify with a particular region or physical habitat. Our preference for mountains, plains, seashores, lakes, big cities, or small towns is probably set at an early age. I became interested in this when I realized that I had no particular preference. The desert, countryside, or city were all the same to me, so long as the humidity was not too high. I thought this might be due to my parents' frequent changes of residence before I was four years old. In America the importance of the environment in which one lives can be seen in the frequency with which immigrants establish themselves in areas that resemble those they have left. Our studies of Stanford students confirmed that those who moved frequently during childhood had trouble identifying with a particular environment and adapted more easily to all environments. Our data did not allow us to determine the most sensitive age, but the study did succeed in showing that a nomadic tendency can be culturally inherited, and that a psychological imprint received while young is difficult to erase later in life. Governments or countries with large nomadic populations (e.g., Gypsies, Bedouins, Berbers, Tuaregs, Pygmies) have difficulty changing their nomadic habits. This poses serious problems for the schooling of their children. Moreover, the freedom of nomadism is fascinating, and if that is the way in which one was raised, it must be very difficult to settle down.

Linguistic Evolution as an Example of Cultural Evolution

Amazingly, linguistic evolution has not been studied much. There is great potential for rigorous quantitative analysis, and research is not very expensive. Interest in language evolution began in the second half of the nineteenth century, by applying the methods of evolutionary trees to the history of language differentiation—especially the Indo-European languages. I have already mentioned that August Schleicher constructed a tree of this family that resembled one based on a recent study. Even though the phenomenon of borrowing words from other languages, especially from neighbors, is well established, the most studied evolutionary trees give the impression that a language changes in ways that are largely independent of changes taking place in other languages. This is a prerequisite for the applicability of tree analysis. We know that languages are often spread over large areas with different varieties (or dialects) developing locally. We are aware that a language changes slightly even within an individual lifetime, but knowledge of old languages is limited, making variation in time somewhat less easy to study than variation in space. Nevertheless, variation in time almost automatically implies variation in space, and the basic rules are the same for both.

What precisely is linguistic variation? There are several aspects. Phonological variation is easily perceived. In any European country, and even in the United States, significant differences in accent occur between north and south, east and west. With a little experience, we can easily guess someone's native region. The pronunciation of words changes in time and space—often significantly.

Another aspect of phonological variation is the richness or poverty of sounds in different languages. Polynesian languages have among the fewest sounds. They have only three vowels: a, i, and u. English is at the opposite extreme with twenty or so vowel sounds (including diphthongs), which are different from those found in all other languages, making the acquisition of English very difficult for foreigners. The speed with which vowel sounds change is particularly astonishing. To paraphrase Voltaire: if the consonants are not very useful for etymological reconstruction, the vowels are completely useless.

Semantic variation is the change of word meanings. For example, the French word *femme* has acquired a second meaning, to include both "woman" and "wife." In Italian, the word *donna* (originally derived from the Latin word *domina* meaning "household master") means "woman," but Italians use *moglie* (from another Latin word, *mulier*) for "wife." Italian also uses *femmina* (from the same root as *femme* in French) to mean only "woman" and not "wife."

Although grammar is the most stable part of a language, it too can change with time. In English, as in French and Italian, the normal word order in a sentence is subject (S), verb (V), and object (O)—or SVO. But all eight possible word orders exist in various languages, even if SVO and SOV languages are much more common. The rarest are OVS and OSV. In the film *Return of the Jedi,* Yoda, the master of the Jedi, uses the OSV style: "Your father he is."

For each of these three modes of linguistic variation (phonological, semantic, and grammatical), change in space is more obvious and easier to study than change in time. We can illustrate on a geographic map the variation a word experiences by drawing a curve delimiting the areas where the word has a particular pronunciation. This curve, which separates one homogeneous region from another, is called an "isogloss." By tracing the isoglosses of many words, we notice that most words display a unique pattern: the limits of pronunciation differ for every word. Where, then, is the region where a single and discrete language or dialect is spoken? The representation of languages with a tree-like monolith, where languages are differentiated from each other in a totally regular manner without being influenced by other languages, is only an approximation.

Five years after Schleicher's work, in 1872, one of his students, Johannes Schmidt, emphasized the importance of local linguistic variation and proposed a theory that in some ways opposed Schleicher's. According to Schmidt, each new form of a word spreads over a geographic area like the waves spreading from a rock thrown into a pond, influencing neighboring speakers to various degrees. This metaphor is particularly apt; it sets itself apart from the model of a tree, which presents completely isolated languages. Can these two views be reconciled?

Theories of biological variation in space, developed in the middle of the twentieth century by several different mathematicians, resulted in very similar models. They have the generic name of "isolation by distance" and show that genes vary randomly in geographic space, following exact rules derived from statistics and probability. The most significant regularity is the relation between genetic distance (calculated from averaging a number of genes) and geographic distance. We have seen that genetic distance increases regularly (but always more slowly) as geographic distance increases, until it reaches a maximum. The shape of the theoretical and empirical curves is determined by two measurable variables: the mutation rate, which increases genetic differences between two places, and the rate of genetic exchange between neighbors due to migration, which tends to increase genetic similarity between them—so these forces are opposed, to a certain extent, and balance each other.

The same mathematical theory can be applied to linguistic evolution: the equivalent of mutation (which produces new forms of genes, or alleles) is innovation, which in linguistics is the generation of new sound, meaning, or grammar. Migration propagates these changes in space. William Wang and I have applied this genetic theory of isolation-by-distance to linguistic variation in Micronesia (1986). One of the most interesting results we have obtained demonstrates that the mutation rate varies greatly for different words. Genes also differ in mutation rates, but less dramatically.

I have already mentioned that there are some words that change very little over time and space, either in their phonology or meaning; they are especially useful for establishing relationships between languages that have been separated for a long time. Unfortunately these words are rare. At the opposite extreme are the highly variable words, the ones that have a high mutation rate. Highly variable genes have a great number of alleles; similarly, highly variable words have a great number of synonyms. They can be found in a thesaurus. For example, there are many synonyms for the word "drunk," undoubtedly because the numerous occasions to use them have produced many jokes. The same is true for the word "penis." Studying the variation of words would certainly offer interesting psychological information.

We must note a significant difference between biological and linguistic mutation. A genetic mutant is generally very similar to the original gene, since one gives rise to another with only a small change. Words vary in more complicated ways. The same root can vary phonologically from language to language and it can also change meaning. One word can have many unrelated senses. One could try to establish greater similarities between genes and words taking into account all of these peculiarities, but it is not clear that this would be useful.

Does the theory of isolation-by-distance destroy the theory of trees? This theory, like Schmidt's, imagines that geographic space is homogeneous. But as we have seen that is not the case. Geographic barriers—seas, oceans, mountains, rivers, and so on—divide the earth into numerous and varied regions. They impede migration and thus hamper the spread of genes and words. By so doing, they create differences between isolated populations. This is what our trees indirectly demonstrate. If the earth's surface were homogeneous and without barriers, a tree would not be useful since the theory of isolation by distance might provide a sufficient and simple description. But if we want a more realistic picture, we must consider the great geographic variability and the richness of historic events which have determined genetic and linguistic patterns. The tree, then, is a useful balance between approximation and reality. Can we sharpen its precision?

Adapting isolation-by-distance theory to a linguistic context allows us to resolve the problem created by Schmidt's theory of waves, and to understand its links with Schleicher's tree-based model. Trees and wave theories show that genetic and linguistic change can be modeled in similar ways, and that it would be useful to explore the similarities and differences between these two types of evolution. In the most basic evolutionary models, four factors interact to cause change: mutation, selection, genetic drift, and migration. Because genetic studies are usually restricted to simply inherited genes and traits (those transmitted by both parents in the way outlined by Mendel), this representation of genetic evolution can ignore a fundamental factor—method of transmission. We have already discussed cultural transmission in general. Regarding linguistic transmission, I will only note that in primitive societies, children learn the language of the family member (mother, sister, etc.) with whom they spend the most time. Thus one would speculate that here transmission is usually vertical, maternal, and uniparental. In more advanced economies, several people may be involved in raising children. At the age when children begin schooling (which differs among cultures and social classes), the teacher has some influence, although friends and classmates do, as well. Here the cultural transmission of language is much more complicated. A child frequently pays inordinate attention to one person (independent of linguistic considerations) and imitates his or her habits, manners, and way of speaking. One person who represents a "role model" can later be replaced by another. Pronunciation is labile until the age of thirteen years or so, after which imitative changes become more rare and potentially unsuccessful. Vocabulary probably comes from the social group of origin but tends to increase throughout life, with exposure to a greater number of people with whom one communicates.

Cultural transmission, which is an important factor in language acquisition, is therefore determined by a series of different transmission mechanisms. The contribution of a true relative can be weak to none, but "adoptive" parents pick up the slack. Each transmitter may contribute something, and the language of an individual may end up as a sort of linguistic mosaic with many different contributions juxtaposed (although a single influence may predominate). After puberty, the cultural product is more or less crystallized. Everyone has his or her accent, which reproduces—with slight variation—those prevalent in the environments where one was raised. Traces of one's earliest learning, covered by later socialization, probably persist, and can reappear in certain circumstances as, for example, when one is tired or placed in a new learning environment.

This analysis, in part autobiographical, is probably not very useful for the scientific literature. But some simplification is necessary and legitimate in order to communicate with a broad audience. We spontaneously correct ourselves, often unconsciously, to use language that will be understood by those with whom we speak. This component of cultural transmission is part of what I have called "concerted" insofar as we make the necessary modifications when we are not being understood.

I have already said that in genetics, as in linguistics, mutations or innovations appear more or less spontaneously in a single individual. They end up comprising part of the linguistic heritage of a population when they are accepted by a significant number of people. Even when change initiated by a single individual is welcomed by the people with whom he is in contact (as a mutation is adopted by the genome), it can take centuries for it to be fully integrated. In genetics, the mutation rate is much lower, and the process of substitution is governed entirely by vertical transmission. Full replacement of the old allele by the new one can take tens or hundreds of thousands of generations. We must, then, understand how and why this increase in frequency occurs.

It is not likely that mutation frequency alone can help a new word diffuse and establish itself in a population—a phenomenon that has been called "mutation pressure" in genetics. But we know that two of the factors in biological evolution—drift and selection—act similarly to affect the rate of substitution of new words. In genetics, drift is the effect of chance. I think that it is difficult to apply the concept of this genetic phenomenon to linguistic change in exactly the same way. Genetic drift depends on the number of individuals in a population and also on the variation of reproductive output of individuals. Not everyone is equal in this respect. Those who reproduce most count the most, although the difference in number of children per parent is usually small. In Europe, only wealthy patrons, like Francesco Sforza, could have thirty or so children. In other countries, a few sultans or chiefs sometimes had hundreds of children. A similar but greatly exaggerated situation applies to language: some people rarely speak, while others talk all the time; variation in the amount of communication is enormous. Moreover, some sources are more highly regarded than others. When these respected people decide to use a new word, it has a greater effect. It is difficult to incorporate such variation in a theory, but it is clear that in linguistic change certain dimensions are more important than in genetics. One can say variation in the prestige of speakers may add greatly to the power of drift. In the past, for example, royalty and nobility determined many changes in language. If they introduced a new word, it was essential to learn it. Today, our language is enriched by radio and television. If one person, accorded great prestige, promotes a new word that is widely accepted, we may say we have an extreme case of drift. But the element of prestige is an unfamiliar and extreme component of drift, and it might seem more appropriate to consider it as a case of cultural selection or of transmission rather than of drift. It is clearly a matter of definition. In some cases, however, the analogy with drift is obvious. Though it is difficult to have precise statistics and valid international comparisons, there is some indication that the U.S. culture is one of the most religious in the world. It is clear that there is a good reason. The religiosity of the U.S. population must come from a strong founders' effect, as shown by the fact that the major contribution to the U.S. culture was from English immigrants in the seventeenth century, who came mostly in search of freedom from religious persecution. American religiosity must be a case of cultural drift.

Imposing biological models on linguistics can present some problems. One is entirely semantic: mostly under the influence of Edward Sapir, linguists use the word "drift" for a different phenomenon. Linguistic drift refers to a trend in a specific direction, noted in a number of similar cases. "The linguistic drift has direction," wrote Sapir. It may be due to a tendency of some linguistic mutations to occur in certain directions. An example is the "Great Vowel Shift." It began in Middle English around the fifteenth century, and is a trend in the change of vowels. Thus, oversimplifying, i → ei → ai → a, a → e → ei; eu → au → ou → uu. Drift in Sapir's sense does not only affect pronunciation but all other aspects of language as well. We will see specific examples later.

This linguistic use of the word "drift" is quite different from the use of the word in genetics, where it has a somewhat opposite meaning. Genetic drift is the effect of chance on gene (allele) frequencies; it is totally devoid of direction, although when allele frequencies of 0 or 100 percent are reached, the process must stop, at least until the lost allele is reintroduced by mutation or immigration from the outside. The use of the word "drift" to indicate the random evolutionary changes of gene frequencies due to chance was suggested by Sewall Wright, who contributed greatly to mathematical work in this area. It was previously called the Hagedoorn effect, after the person who first described it. Another famous mathematical geneticist who also gave much impetus to the theory of drift, Motoo Kimura, suggested the phrase "random genetic drift" might be more precise. The word "drift" is used in linguistics and in other disciplines, like physics, to define systematic effects, as opposed to chance effects.

Selection also works differently in the evolution of language than in biological evolution. It is very rare, of course, that a new word will increase the survival, or reproduction, of the people who use it. Instead, it is a matter of cultural selection: a word, a pronunciation, or a rule pleases us because the word is shorter, easier to pronounce, more elegant, et cetera, or because it is recommended by someone we respect. We could adopt the language of kings, or assume the accent of Oxford dons, but the opposite phenomenon is also popular. Slang seems more efficient to us, because it is richer in emotional undertones. Likewise, a well-educated person might prefer to use a vulgar word, precisely because its use is shocking and therefore powerful. We said that lower social classes often tend to imitate higher classes and vice versa. This tends to create cyclic behavior. In England, mostly in the higher social classes, it was considered more elegant to use words from Latin or its Romance descendants than to use Anglo-Saxon words—for example "serviette" instead of "napkin." Today Anglo-Saxon words have acquired a new dignity and the opposite trend has begun.

So far, we have neglected the fourth factor—migration, both of individuals and of words. Today, words can spread without the movement of people. Once, they spread only with the people who spoke them. We often think that each ethnic group is totally endogamous—that is, marriages are limited to specific social classes or geographic neighborhoods. In reality, there is almost always some genetic exchange between geographic, ethnic, or socio-economic groups. The frequency with which a spouse (usually the wife) comes from another tribe or another village is highly variable, and ranges mostly between 5 and 50 percent. Linguist Joseph Greenberg has observed that transplanted spouses are likely to bring linguistic novelties with them. An interesting rule, which I learned from Claude Hagège, is that island populations exhibit linguistic inertia: it appears that their language ceases almost entirely to evolve. This has happened in Iceland, which was settled by Norwegians in the ninth century A.D. Modern Icelandic is very similar to ancient Norse, and Icelandic speakers are easily able to read the great epics, the sagas that date from the colony's founding or earlier. Outside contact diminished greatly and virtually stopped after the eleventh century; linguistic novelties then ceased to arrive. The rarity of migrants was like a lack of mutation. Without new material, evolution stops. Almost the entire Icelandic population used to meet yearly when Parliament (the first in European history) met; this probably helped avoid excessive local differentiation within the island, and contributed to slowing down evolution.

Another example is Sardinia—the most isolated Mediterranean island—though it has a longer history than Iceland. The coast was less isolated than the interior, where mountains hindered even the Romans. Besides its insularity, every aspect of Sardinia's geography favors the conservation of local culture and language. As a result, some words and endings remained closer to Latin in Sardinia than in continental Italy.

We could not leave this subject without mentioning the most interesting aspect of linguistic evolution—lexical diffusion, whose importance was demonstrated by William Wang. Lexical diffusion does not refer to the way an innovation spreads from one person to another, but to the effect that the change in one word may have on other words in one person's vocabulary. This is especially important, because it also tells us about the working of the brain, which seems to operate from a set of rules. Although each language preserves many grammatical, phonological, and syntactical irregularities, there is a tendency for homogenization and extension of rules. English verbs are in the process of becoming more regular as time passes. Another example is the differentiation of verbs and nouns by the position of the accent: the word "present" is a noun if the accent is on the first syllable, and a verb if on the second syllable. In 1570 there were only three examples (outlaw, rebel, record). Between 1582 to 1934 they grew steadily from 8 to 150.

In general, lexical diffusion means that changes which, for various reasons, are made in one word are often extended to words which are in some way (usually phonologically or grammatically) related to the first. As always, the phenomenon most probably presents itself at first in one or few persons, but then spreads to other people; there is thus a double diffusion, to related words within a person, and to different persons.

Lexical diffusion may be very general. Some linguists may be shocked because the idea was never presented to them, but it seems reasonable to consider as an example of lexical diffusion the classic correspondence of sounds called Grimm's Law, which explains that the letters *p, t,* and *k* of ancient languages like Sanskrit, Greek, and Latin have usually become *f, th,* and *h,* respectively, in English and *f, d,* and *h* in German (e.g., *pater* in Latin, "father" in English, *vater* pronounced "fater" in German). In English, spelling rules were fixed before the Renaissance and an important shift in vowel pronunciation, the "Great Vowel Shift" mentioned above, began at the end of the Middle Ages. A difficult English orthography developed as a result. For example, before the Great Vowel Shift the words "mine," "fine," and "thine" were pronounced phonetically, as they were written—in other words, the *i* was pronounced as in Italian and the *e* was not silent. Then the pronunciation of *i* became *ii,* then *ei* and in modern English *ai.* In some parts of England, especially those far from London, the old pronunciations have been preserved. Elsewhere, other pronunciations are found like *a* or *oi.* From an evolutionary point of view, some are more advanced, because they probably already passed through the forms of *ei* and *ai.* As change tends to be cyclic in the case of vowels, the original pronunciation may return to favor. One reason for this is that the space in which phonological variation can take place is limited, and repetition unavoidable. Cycles form because there are preferential patterns of change.

In Brazil, the old form of Portuguese, which pronounced the *t* at the end of words, as in the English "dent" and "president," persists in the south, but has been replaced with the pronunciation *tch* in the north. The *n* that regularly precedes *st* or *sc* in Latin has been kept in most European languages, but has been dropped in a large number of Italian words. Thus "institute, instance, inscription" have become *istituto, istanza, iscrizione.* When *n* was useful to distinguish two meanings it disappeared in one and remained in the other: *ispirare* (inspire), but *inspirare* (breathe in).

The extension of changes to similar meanings or sounds is the fundamental characteristic of lexical diffusion. It occurs with remarkable speed, sometimes perhaps within a single generation, a clear indication that our brain uses rules for speaking. The need for the human brain to function according to rules must be based in specific neurological structures. Some pathological conditions produce dyslexias, which seem to affect the relevant part of the brain. One of these dyslexias that appears to be inherited as a regular gene in a particular family is being studied from a genetic perspective. This single gene apparently affects the capacity to apply grammatical rules, like the formation of plurals. In affected members of this family, only words for which the singular and plural were learned separately are used correctly. Perhaps one could consider an example of lexical diffusion the increasing rarity of the use of the subjunctive in English (and in Italian). Grammar appears to require certain neurological centers, and genetic defects or cerebral trauma that damage them can interfere with the application of grammatical rules. The same defects can sometimes spread without any obvious pathological cause. Such observations reveal mechanisms, previously unknown, that help us use words in a coherent fashion. Lexical diffusion must depend on equivalent mechanisms that facilitate linguistic function.

Humanity's Future

Rest assured that my intention in this section is more modest than the title above suggests. From a genetic perspective our human future is not terribly interesting—our species will probably not evolve much more. In any case, it will not evolve as rapidly as it has so far. Cultural development has effectively slowed biological evolution. Natural selection, by acting on fertility and mortality, has been the greatest evolutionary factor in human biology. But progress in medicine has virtually eliminated pre-reproductive mortality to such an extent that demographic growth must now be sharply curtailed to prevent serious overpopulation. If pre-reproductive mortality were reduced to zero, everybody married, and each family had two children, there would be no natural selection. Because of the large, continuously increasing size of human populations, the other cause of evolution, genetic drift, is almost completely frozen. We now consider mutations dangerous, since they involve changes to DNA that are on average harmful. Why not stop mutation if it becomes possible to do so? Human biological evolution would then stop completely, as long as we avoid the mistake of voluntary evolution by artificially modifying genes. Fortunately, the likelihood of genetically engineered humans is still almost nonexistent, and we do not yet have to worry about some arrogant fool attempting to create an "improved race." Naturally, special resolutions, such as those governing nuclear weapons technology, will have to be implemented to avoid a nightmarish distant future.

One important genetic change is, however, currently occurring via migrations that produce increased mixing of populations. If, as is likely, the process continues, genetic differences between groups will diminish. But the overall global diversity will not change, and differences between individuals of the same population will increase. There will thus be even fewer reasons for racism, which is a good thing.

It is not strictly correct, however, to say that global variation will remain unchanged. At the moment, different ethnic groups have different reproductive rates. Europeans are largely at a standstill while populations in many developing countries are exploding; thus blonds and light-skinned people will decline in relative frequency. But even those who do not worry about the excessive reproduction of the human species will soon learn that the current population boom cannot continue beyond what Earth's resources can support. This means that it must stop in a few decades.

It is clear that the rate of cultural change will continue to increase in the future. Communication forms the basis of cultural change, and we are currently in the midst of a communications revolution. Where will it take us? Computers have acted to some extent as an extension of our brains and have greatly increased our capacity for numeric calculations. Artificial intelligence is currently extending computer applications in new directions.

As it was during the Paleolithic, human communication is limited, despite technology, by linguistic barriers. Computers have so far been unable to translate human languages automatically. Difficult as it is to resolve the problem, it is only a matter of time before we have automatic translation of a reasonable quality. Perhaps we will be able to learn to speak in a less ambiguous way, allowing the computer to understand and translate our thoughts with fewer errors. It seems incredible in light of recent progress that computers still have this difficulty. Certainly our expressions are often ambiguous. Sometimes we confuse each other on purpose. Decreasing language ambiguity may reduce the chances of writing good poetry, and perhaps a remedy could be found for that, but it should force even politicians to think clearly and productively for their constituencies, not just for their reelection or profit.

Nevertheless, automatic translations are not the solution to all our problems. Communication is certainly essential, but only as a first step. It will be necessary, for example, to be more successful in spreading the necessary moral values to the whole world. Is the amount of deception, hatred, exploitation, and unrestrained selfishness we observe in almost every society inevitable? We need not be too pessimistic and should admit that people do not always display their worst qualities. But it would be valuable to learn exactly the conditions that elicit these destructive tendencies, in order to systematically prevent them. Overpopulation and extreme competition for valuable resources undoubtedly contribute. Our aptitude for social engineering is limited, although we must become more serious about work in this area, so as to end—or at least reduce—major social ills such as poverty, ignorance, population growth, racism, drug addiction, crime, and other social epidemic and endemic diseases that afflict us. Our efforts in this regard can be helped by studying cultural transmission and the forces of conservatism that hinder useful innovations, as well as the danger posed by promoting and accepting great changes too soon.

BIBLIOGRAPHY

Ammerman, A. J. and Cavalli-Sforza, L. L. (1984) *The Neolithic Transition and the Genetics of Populations in Europe* (Princeton University Press, Princeton, N.J.).

Anthony, D. W. (1995) Horse, wagon and chariot: Indo-European languages and archaeology. *Antiquity* 69: 554–65.

Bailey, N. J. (1957) *The Mathematical Theory of Epidemics* (Hafner, New York).

Barbujani, G. and Sokal, R. R. (1990) Zones of sharp genetic change in Europe are also linguistic boundaries. *Proc. Natl. Acad. Sci.* 87(5): 1816–19.

Barbujani, G., Magagni, A., Minch, E. and Cavalli-Sforza, L. L. (1997) An apportionment of human DNA diversity. *Proc. Natl. Acad. Sci.* 94: 4516–19.

Barrantes, R., Smouse, P. E., Mohrenweiser, H. W., Gershowitz, H., Azofeifa, J., Arias, T. D. and Neel, J. F. (1990) Microevolution in lower Central America: Genetic characterization of the Chibcha-speaking groups of Costa Rica and Panama, and a consensus taxonomy based on genetic and linguistic affinity. *Am. J. Hum. Genet.* 46(1): 63-84.

Biraben, N.-J. (1980) An essay concerning mankind's evolution, *Population* 4: 1–13.

Bowcock, A. M., Kidd, J. R., Mountain, J. L., Hebert, J. M., Carotenuto, K., Kidd, K. K. and Cavalli-Sforza, L. L. (1991) Drift, admixture, and selection in human evolution: A study with DNA polymorphisms. *Proc. Natl. Acad. Sci.* 88: 839–43.

Bowcock, A. M., Ruiz-Linares, A., Tomfohrde, J., Minch, E., Kidd, J. R. and Cavalli-Sforza, L. L. (1994) High resolution of human evolutionary trees with polymorphic microsatellites. *Nature* 388: 455–57.

Cann, R. L., Stoneking, M. and Wilson, A. C. (1987) Mitochondrial DNA and human evolution. *Nature* 325: 31–36.

Cappello, N., Rendine, S., Griffo, R. M., Mameli, G. E., Succa, V., Vona, G. and Piazza, A. (1996) Genetic analysis of Sardinia. *Ann. Hum. Genet.* 60: 125–41.

Cavalli-Sforza, L. L. (1963) The distribution of migration distances, models and applications to genetics, in Human displacements: measurement, methodological aspects, ed. Sutter, J. (Éditions Sciences Humaines, Monaco), pp. 139–58.

——— (1986) *African Pygmies* (Academic Press, Orlando, Fla.).

——— (1998) The DNA revolution in population genetics. *Trends in Genetics.* 14(2): 60–65.

Cavalli-Sforza, L. L. and Cavalli-Sforza, F. (1995) *The Great Human Diasporas* (Addison-Wesley, Menlo Park, Calif.).

Cavalli-Sforza, L. L. and Edwards, A. W. F. (1964) Analysis of human evolution. *Proc. 11th Int. Congr. Genet.* 2: 923–33.

——— (1967) Phylogenetic analysis: Models and estimation procedures. *Am. J. Hum. Genet.* 19: 223–57.

Cavalli-Sforza, L. L. and Feldman, M. (1981) *Cultural Transmission and Evolution, A Quantitative Approach* (Princeton University Press, Princeton, N.J.).

Cavalli-Sforza, L. L., Feldman, M. W., Chen, K. H. and Dornbusch, S. M. (1982) Theory and observation in cultural transmission. *Science* 218: 19–27.

Cavalli-Sforza, L. L., Menozzi, P. and Piazza, A. (1993) Demic expansions and human evolution. *Science* 259: 639–46.

——— (1994) The History and Geography of Human Genes (Princeton University Press, Princeton, N.J.).

Cavalli-Sforza, L. L., Minch, E. and Mountain, J. (1992) Coevolution of genes and language revisited. *Proc. Natl. Acad. Sci.* 89: 5620–24.

Cavalli-Sforza, L. L. and Piazza, A. (1975) Analysis of evolution: evolutionary rates, independence and treeness. *Theor. Popul. Biol.* 8: 127–65.

Cavalli-Sforza, L. L., Piazza, A., Menozzi, P. and Mountain, J. L. (1988) Reconstruction of human evolution: Bringing together genetic, archaeological, and linguistic data. *Proc. Natl. Acad. Sci.* 85: 6002–06.

Cavalli-Sforza, L. L. and Wang, W.S.-Y. (1986) Spatial distance and lexical replacement. *Language* 62: 38–55.

Coale, A. J. (1974) The history of the human population. *Sci. Amer.* 231(3): 40–51.

Dolgopolsky, A. B. (1988) The Indo-European homeland and lexical contacts of Proto-Indo-European with other languages. *Mediterr. Lang. Rev.* (Harrassowitz) 3: 7–31.

Durham, W. H. (1991) *Coevolution: genes, culture, and human diversity* (Stanford University Press, Stanford, Calif.).

Dyen, I., Kruskal, J. G. and Black, P. (1992) An Indo-European classification: a lexicostatistical experiment. *Transactions of the Amer. Philosph. Society* 82: Part 5 (American Philosophical Society, Philadelphia, Pa.).

Efron, B. (1982) *The Jackknife, Bootstrap, and Other Resampling Plans* (Society for Industrial and Applied Mathematics, Philadelphia, Pa.).

Felsenstein, J. (1973) Maximum-likelihood estimation of evolutionary trees from continuous characters. *Am. J. Hum. Genet.* 25: 471–92.

——— (1985) Confidence limits on phylogenies: an approach using the bootstrap. *Evolution* 29: 783–91.

Gamkrelidze, T. V. and Ivanov, V. V. (1990) The Early History of Languages. *Sci. Amer.* 263(3): 100–16.

Gimbutas, M. (1970) *Proto-Indo-European culture: The Kurgan culture during the fifth, fourth and third millennia B.C. in Indo-European and Indo-Europeans,* ed. Cardona, G. R., Hoenigswald, H. M. and Senn, A. (University of Pennsylvania Press, Philadelphia, Pa.) pp. 155–95.

——— (1991) *The Civilization of the Goddess* (Harper, San Francisco, Calif.).

Goldstein, D. B., Ruiz-Linares, A., Cavalli-Sforza, L. L. and Feldman, M. W. (1995) Genetic absolute dating based on microsatellites and the origin of modern humans. *Proc. Natl. Acad. Sci.* 92: 6723–27.

Greenberg, J. H. (1987) *Language in the Americas* (Stanford University Press, Stanford, Calif.).

Greenberg, J. H., Turner II, C. G. and Zegura, S. L. (1986) The settlement of the Americas: a comparison of the linguistic, dental, and genetic evidence. *Curr. Anthropol.* 27(5): 477–97.

Guglielmino, C. R., Viganotti, C., Hewlett, B. and Cavalli-Sforza, L. L. (1995) Cultural variation in Africa: Role of mechanisms of transmission and adaptation. *Proc. Natl. Acad. Sci.* 92: 7585–89.

Hewlett, B. S. and Cavalli-Sforza, L. L. (1986) Cultural transmission among the Aka pygmies. *Am. Anthropol.* 88: 922–34.

Hiernaux, J. (1985) *The People of Africa* (Scribner's, New York).

Horai, S., Hayasaka, K., Kondo, R., Tsugane, K. and Takahata, N. (1995) Recent African origin of modern humans revealed by complete sequences of hominoid mitochondrial DNAs. *Proc. Natl. Acad. Sci.* 92: 523–26.

Howells, W. W. (1973) Cranial variation in man: a study by multivariate analysis of patterns of difference among recent human populations. *Peabody Mus. Archaeol. Ethnol. Harv. Univ.* 67: 1–259.

——— (1989) Skull shapes and the map: craniometric analyses in the dispersion of modern *Homo. Pap. Peabody Mus. Archaeol. Ethnol. Harv. Univ.* 79: 1–189.

Kimura, M. and Weiss, G. H. (1964) The stepping-stone model of population structure and the decrease of genetic correlation with distance. *Genetics* 49: 561–76.

Kruskal, J. B. (1971) Multi-dimensional scaling in archaeology: time is not the only dimension in Mathematics in the Archaeological and Historical Sciences, ed. Hodson, F. R., Kendall, D. G. and Tautu, P. (Edinburgh University Press, Edinburgh), pp. 119–32.

Kruskal, J. B., Dyen, I. and Black, P. (1971) The vocabulary and method of reconstructing language trees: innovations and large scale applications, ibid. pp. 361–80.

Le Bras, H. and Todd, E. (1981) *L'invention de la France: Atlas, Anthropologique et Politique* (Livre de Poche, Hachette, Paris).

Li, J., Underhill, P. A., Doctor, V., Davis, R. W., Shen, P., Cavalli-Sforza, L. L. and Oefner, P. (1999) Distribution of haplotypes from a chromosome 21 region distinguishes multiple prehistoric human migrations. *Proc. Natl. Acad. Sci.* (USA) 96. 3796–800.

Malécot, G. (1948) *Les Mathématiques de l'Hérédité* (Masson, Paris).

——— (1966) *Probabilité et Hérédité* (Presses Universitaires de France, Paris).

Mallory, J. P. (1989) *In Search of the Indo-Europeans: Language, Archaeology and Myth* (Thames and Hudson, London).

Menozzi, P., Piazza, A., and Cavalli-Sforza, L. L. (1978) Synthetic maps of human gene frequencies in Europe. *Science* 201: 786–92.

Morton, N. E., Yee, S. and Lew, R. (1971) Bioassay of kinship. *Biometrics* 27(1): 256.

Mountain, J. L. and Cavalli-Sforza, L. L. (1994) Inference of human evolution through cladistic analysis of nuclear DNA restriction polymorphisms. *Proc. Natl. Acad. Sci.* 91: 6515–19.

Mountain, J. L., Lin, A. A., Bowcock, A. M. and Cavalli-Sforza, L. L. (1992) Evolution of modern humans: Evidence from nuclear DNA polymorphisms. *Phil. Trans. R. Soc. Lond.* (B) 377: 159–65.

Mourant, A. E. (1954) *The Distribution of the Human Blood Groups* (Blackwell Scientific, Oxford).

Murdock, G. P. (1967) *Ethnographic Atlas* (University of Pittsburgh Press, Pittsburgh, Pa.).

Nei, M. (1987) *Molecular Evolutionary Genetics* (Columbia University Press, New York).

Penny, D., Watson, E. E. and Steel, M. A. (1993) Trees from genes and languages are very similar. *Sist. Biol.* 42: 382–84.

Piazza, A., Minch, E. and Cavalli-Sforza, L. L. Unpublished manuscript on the tree of sixty-three Indo-European languages.

Piazza A., Rendine, S., Minch, E., Menozzi, P., Mountain, J. and Cavalli-Sforza, L. L. (1995) Genetics and the origin of European languages. *Proc. Natl. Acad. Sci.* 92: 5836–40.

Poloni, E. S., Excoffier, L., Mountain, J. L., Langaney, A. and Cavalli-Sforza, L. L. (1995) Nuclear DNA polymorphism in a Mandenka population from Senegal: comparison with eight other human populations. *Ann. Hum. Genet.* 59: 43–61.

Quintana-Murci, L., Semino, O., Bandelt, H-J., Passarino, G., McElreavey, K. and Santachiara-Benerecetti, A. S. (1999) Genetic evidence of an early exit from Africa through eastern Africa. *Nat. Genet.* 23: 437–441.

Rendine, S., Piazza, A. and Cavalli-Sforza, L. L. (1986) Simulation and Separation by Principal Components of Multiple Demic Expansions in Europe. *American Naturalist* 128: 681–706.

——— (1989) The origins of Indo-European languages. *Sci. Amer.* 261(4): 106–14.

Renfrew, C. (1987) *Archaeology and Language: The Puzzle of Indo-European Origins* (Jonathan Cape, London).

Ruhlen, M. (1987) *A Guide to the World's Languages* (Stanford University Press, Stanford, Calif.).

——— (1991) Postscript in *A Guide to the World's Languages* (Stanford University Press, Stanford, Calif.), pp. 379–407.

Saitou, N. and Nei, M. (1987) The neighbour-joining method: a new method for reconstructing phylogenetic trees. *Mol. Biol. Evol.* 4(4): 406–25.

Seielstad, M. T., Minch, E. and Cavalli-Sforza, L. L. (1998) Genetic evidence for a higher female migration rate in humans. *Nat. Genet.* 20: 278–280.

Semino, O., Passarino, G., Brega, A., Fellows, M. and Santachiara-Benerecetti, A. S. (1996) A view of the Neolithic diffusion in Europe through two Y-chromosome-specific markers. *Am. J. Hum. Genet.* 59: 964–8.

Sokal, R. R., Harding, R. M. and Oden, N. L. (1989) Spatial patterns of human gene frequencies in Europe. *Am. J. Phys. Anthropol.* 80: 267–94.

Sokal, R. R. and Michener, C. D. (1958) A statistical method for evaluating systematic relationship. *Univ. Kansas Sci. Bull.* 38: 1409–38.

Stigler, S. M. (1986) *The History of Statistics* (Harvard University Press, Cambridge, Mass.).

Sutter, J. (1958) Recherches sur les effets de la consanguinité chez l'homme. *Bio. Med.* 47: 463–60.

Tobias, P. V. (1978) *The Bushmen: San Hunters and Herders of South Africa* (Human and Rousseau, Cape Town).

Todd, E. (1990) *L'invention de l'Europe* (Éditions de Seuil, Paris).

Turner II, C. G. (1989) Teeth and prehistory in Asia. *Sci. Amer.* 260(2): 88–96.

Underhill, P. A., Jin, L., Lin, A. A., Medhi, S. Q., Jenkins, T., Vollrath, D., Davis, R. W., Cavalli-Sforza, L. L. and Oefner, P. J. (1997) Detection of numerous Y chromosome biallelic polymorphisms by denaturing high-performance liquid chromatography. *Genome Res.* 7: 996–1005.

Underhill, P. A., Jin, L., Zemans, R., Oefner, P. and Cavalli-Sforza, L. L. (1996) A pre-Columbian Y chromosome-specific transition and its implications for human evolutionary history. *Proc. Natl. Acad. Sci.* 93: 196–200.

Warnow, T. (1997) Mathematical approaches to comparative linguistics. *Proc. Natl. Acad. Sci.* 94: 6585–6590.

Wolf, A. P. (1980) Marriage and Adoption in China, 1854–1945 (Stanford University Press, Stanford, Calif.).

Zei, G., Astolfi, P. and Jayakar, S. D. (1981) Correlation between father's age and husband's age: a case of imprinting? *J. Biosoc. Sci.* 13: 409–18.

Zei, G., Barbujani, G., Lisa, A., Fiorani, O., Menozzi, P., Siri, E. and Cavalli-Sforza, L. L. (1993) Barriers to gene flow estimated by surname distribution in Italy. *Ann. Hum. Genet.* 57: 123–140.

Zuckerkandl, E. (1965) The evolution of hemoglobin. *Sci. Amer.* 212: 110–18.

INDEX